Tropical Forests

Tropical Forests

Regional Paths of Destruction
and Regeneration in the Late
Twentieth Century

Thomas K. Rudel

Columbia University Press   New York

Columbia University Press
Publishers Since 1893
New York Chichester, West Sussex
Copyright © 2005 Columbia University Press
All rights reserved

Library of Congress Cataloging-in-Publication Data
Rudel, Thomas K.
   Tropical forests : regional paths of destruction and regeneration in the late twentieth
century / Thomas K. Rudel.
      p. cm.
   Includes bibliographical references (p.).
   ISBN 0-231-13194-1 (cloth : alk. paper) — ISBN 0-231-13195-X (pbk. : alk. paper)
   1. Deforestation — Tropics.   2. Rain forest conservation.   I. Title

SD418.3.T76T763 2005
333.75′137′0913 — dc22   2005041419

∞

Columbia University Press books are printed on permanent and durable acid-free paper.

Printed in the United States of America

c 10 9 8 7 6 5 4 3 2 1
p 10 9 8 7 6 5 4 3 2 1

In memory of William E. Rudel, 1916–1991

Contents

List of Figures

List of Tables

Preface

Concern for tropical forests usually comes from direct experience. Most of the people around the globe who work to conserve tropical forests do so with a particular forest in mind. It may be the forest over the hill behind their house, or it may be a forest thousands of miles away that they visited as part of a study tour while in college. First-hand knowledge about forests helps people conserve them. In particular, it sharpens our understanding of the practical issues involved in maintaining a forest used by many people. It also brings with it a liability. The images of the forests that we know are usually so compelling that it is very easy to generalize the attributes of these familiar forests to other forests that we do not know. An institutional arrangement that conserves forests and provides economic benefits for local people in one place becomes a solution to dilemmas of environmental conservation in other places. Unfortunately, the circumstances that imperil forests and impoverish people vary considerably across rain forest regions, so solutions for one place rarely apply to other places.

When I think of tropical rain forests, I think of forests at the eastern base of the Andes in Ecuador. As I began to research the problems of tropical deforestation during the 1980s and read the research reports about tropical forest destruction in other countries, I became aware of differences between conditions in the Ecuadorian place that I knew so well and conditions in other places. These differences in conditions call for careful contextualization in research reports so that readers can gauge the degree to which the findings from one place might apply to other places. Very few articles address these questions of scope, so readers are left guessing. Presumably, the conditions attending the destruction of tropical forests in the place under study might apply to all other places in the tropical realm.

Evidence for geographic differences in forest-cover change abounds. Cattle ranching plays an important role in the Americas but not elsewhere. Logging has figured centrally in forest destruction in Southeast Asia, but it seems less important elsewhere. Forests differ tremendously in size between countries in ways that would seem to make them more or less costly to exploit, but few analysts have noted these differences. Of course cross-national quantitative analyses would capture some geographic differences, but these types of analyses remained so disembodied and decontextualized that they shed little insight on the forces driving forest declines and gains. These circumstances call out for a full-bodied comparative analysis of the assembled case studies, not because we need to solve another academic puzzle, but because appropriately contextualized descriptions of tropical forest-cover change should contribute to policies for tropical forest conservation that fit a particular time and place. The last chapter of this book takes up the challenge of crafting geographically explicit conservation policies by identifying policies for regions with large forests and policies for regions with small forests.

My concern with tropical flora and fauna began more than 30 years ago, when, as a U.S. Peace Corps volunteer fresh out of college, I helped a group of poor, landless families from the Andean highlands "carve" farms and new lives out of the Amazon rain forest in eastern Ecuador. I was proud to be helping poor people start new, more prosperous lives and sad to see these new lives achieved amid the wreckage of a recently felled rain forest. Helping poor people achieve a small measure of economic security by mowing down a rain forest underlined for me in a compelling, personal way the conflicts between environment and society. I am still struggling with these conflicts, both in a personal and in a professional way. They underscore the importance of trying to achieve some kind of sustainable development. With this book, I hope that I can make a small contribution toward that end.

Acknowledgments

During the course of researching and writing this book I accumulated debts of gratitude to some organizations and many people. Funding first from the National Science Foundation (SES 91–23569) and later from the Forest Resource Assessment 2000 program, under the direction of Peter Holmgren, at the Food and Agricultural Organization (FAO) of the United Nations provided the financial resources necessary to demonstrate proof of concept, that a meta-analysis of the tropical deforestation literature could actually tell us new things about forest cover change in the tropics. The library staff at Rutgers University, in particular Dean Meister, responded to my never-ending requests for articles from out-of-the-way sources with dispatch and good cheer. Diane Bates, Kevin Flesher, Jill Roper, and Sandra Baptista, then graduate students in ecology, geography, and sociology at Rutgers, put in long hours compiling references and then reading the burgeoning literature on the human-induced changes in tropical forests. Their work demonstrated that it was possible to find and acquire all of the case studies of tropical forest cover change. By reading and coding all of these studies during the initial phase of the research, they highlighted for me the challenges of trying to devise a single coding scheme to be used by multiple readers for articles written by both natural and social scientists.

Once I began the meta-analysis, other people helped me at crucial junctures. Kriss Drass (now deceased) answered my many questions about qualitative comparative analysis whenever I asked. Mike Siegel prepared the maps of each regional rain forest and explained to me, with great patience, why it was difficult to create complex maps for small spaces in books! Kojo Amanor and Jill Belsky provided important biographical details for several of the people profiled in the text. Lena Raberg did meticulous work in preparing the appendix that lists all of the case studies and shows how I coded them. Bill McConnell and Eric Keys provided a platform for assessing the validity of this approach to studying forest-cover change by allowing me to present my findings at three Land Use and Cover Change (LUCC) workshops: on Latin America in Tempe, Arizona, in November 2002, on Africa in Wageningen, Netherlands, in May 2003, and on Asia in Honolulu, Hawaii, in November 2003. The participants in each workshop had carried out case studies in their respective regions. I owe them a large debt of gratitude for their comments on the meta-analyses; their critical comments prompted me to undertake further meta-analyses that without a doubt improved the final product. William Sunderlin and Sven Wunder read an entire draft of the manuscript at an early stage. With their field experience on all three continents (Africa, Asia, and Latin America), William and Sven made a series of trenchant comments that I have tried to address in revising the book. The anonymous reviewers at Columbia University Press made a series of helpful comments, pointing out omissions and confusions. One of them suggested an improved title for the book that I have adopted. I thank them for their investment of time and intellectual energy in this book. Robin Smith at Columbia University Press provided a judicious mix of encouragement and critical comments that kept me moving on the manuscript.

Susan Golbeck and Dan Rudel provided the never-ending series of distractions that make home life both a respite and a refuge from the sometimes frustrating efforts to see patterns and say what I mean about them. Finally, I would like to dedicate this book to my father, as a belated and inevitably inadequate way of thanking him for introducing me to the outdoors and helping me cultivate a love for it. He set me on a life course that I am still pursuing.

Tropical Forests

1
Introduction

If the tropical rain forests are all cut down, we will never know what we have lost.... Species will go extinct before they are ever discovered.
A. FORSYTH AND K. MIYATA, *Tropical Nature* (1983:211)

Paths of Destruction and the Promise of Regional Analyses

Fears about the imminent extinction of large numbers of plants and animals have prompted an outpouring of concern and analysis about tropical deforestation during the past two decades. Despite the emergence of a social movement devoted to reducing deforestation, and the publication of hundreds of studies that analyzed its causes, the destruction of tropical rain forests did not appear to slow down much, if at all, during the 1990s (Stokstad, 2001).[1] Brazil provides a prominent case in point. After reaching high rates in the late 1980s, deforestation rates in the Brazilian Amazon declined during the early 1990s, only to increase to a very high rate in 1995. After a subsequent decline, rates of deforestation have increased steadily over the past 6 years. They now approximate the high rates of deforestation observed during the late 1980s (Instituto Nacional de Pesquisas Espaciales, 2004). In recent years, extensive wild-

fires on cutover lands in the tropics have added a new dimension, pol-
luted air, to the environmental destruction that has accompanied tropical
deforestation.

The pessimistic language and revised policy positions of people active
in the environmental movement reflect the continued loss of forests. The
titles of recently published books on tropical rain forests convey a somber message: "Requiem for Nature" (Terborgh, 1999), "The Last Frontier Forests: Ecosystems and Economies on the Edge" (Bryant, Nielsen,
and Tangley, 1997), and "Last Stand: Protected Areas and the Defense
of Tropical Diversity" (Kramer, Van Schaik, and Johnson, 1997). Some
environmentalists have begun to refer to the earth as a "planet of weeds."
John Terborgh (1999:20), a prominent ecologist, summed up the prevailing sentiment in the late 1990s: "Short of radical changes in government
policy in country after country, all unprotected tropical forests appear
doomed to destruction within thirty to fifty years." Because conservationists do not have the political power to save most tropical rain forests,
some analysts have begun to argue that conservationists should concentrate their efforts on saving the few forests in remote interiors that are not
in danger of imminent destruction (Pearce, 1998).

Point of view makes a difference in observers' assessments. The
global prospects for conservation elicit the most frequent expressions
of despair because the magnitude of the task seems so overwhelming.
The local prospects for conservation sometimes seem more promising because observers see possibilities for crafting conservation policies that build on preexisting trends in the local society and economy.
These supporting conditions are difficult to see at the global scale,
but they are crucial to the success of conservation efforts. For this
reason, when environmentalists assess the prospects for conservation
in particular places, the task seems more manageable, and observers
express more optimism (Mittermeier et al., 2000:23). Although these
local-level analyses have the virtue of pointing to clear policy directions, they often lack a sense of proportion. The promise of ecological restoration in a small upland watershed in Mindinao (Stone and
D'Andrea, 2001:59–79) does not somehow offset the rapid destruction
of old growth forests in the much larger lowlands of Indonesia's outer
islands (Jepson et al., 2001).

Between the pessimism of the global and the sometimes facile optimism of the local lies a third way, a regional, meso-level approach to

questions that pays analytical as well as emotional dividends. Only at this level of aggregation can analysts combine assessments of driving forces from the larger context with the local features of a regional situation in a way that reveals the possibilities for conservation in a place. For this reason, Meyer and Turner (1992:57) recommend that researchers "seek a middle range between the global and the local" in assessing the forces that drive land use conversion in the tropics.

The analytical advantages of a regional focus stem in part from the number of important factors in tropical deforestation that are regional in extent. Perhaps of greatest importance, the forests themselves, founded on climatic conditions, have regional dimensions. The coastal forests of West Africa extend across thirteen countries; the Amazon basin forests stretch across nine countries. The agricultural economies of tropical regions follow suit. Smallholders in West Africa cultivate cacao on freshly cleared forest lands in at least nine of the thirteen countries in the region, so rising and falling prices for cacao have a regional effect. Similar arguments could be made about coffee in Central America and timber in Southeast Asia.

Although the regional extent of rain forests and related land uses make it easier to conceptualize processes of land use conversion at meso-levels of aggregation, it would be a mistake to restrict our attention to the regional scale. The individual and institutional actors responsible for tropical deforestation range across many scales in social life, from local landowners to global markets. As numerous analysts have noted (Watson, 1978; Turner, 1999), compelling accounts of deforestation draw on data from several different scales. These studies describe the local circumstances surrounding forest destruction as well as the regional and global contexts that shape and constrain land clearing (Little, 2000:105). Meso-scale analyses are particularly fruitful in this regard because analysts can scale both *up* from the local level and *down* from the global level (McConnell and Moran, 2000). *Tropical Forests* tries to deliver on this analytical promise. It uses a comparative historical method to outline regional trajectories of land-cover change in the tropics during the past three decades. A clear delineation of these historical paths and associated turning points should make it easier for environmentally concerned people to construct regional conservation strategies that acknowledge and build on preexisting trajectories of forest cover change (Kaimowitz, 2002).

Evaluating Information About Tropical Deforestation

Do the hundreds of studies and reports about tropical deforestation provide descriptions of regional processes and guidelines for regional conservation strategies? Unfortunately, they do not. Both the agencies' reports and scientists' studies suggest that regional characterizations of deforestation processes might be informative, but problems with both sources of information make it impossible to use them as more than points of departure for a more extended study of regional variations in deforestation processes. The strengths and weaknesses of these sources of information and their indicators of forest cover change are outlined in the following sections.

Agency Reports About Tropical Deforestation

Since the early 1980s, reports from the Forestry Division of the Food and Agricultural Organization of the United Nations (FAO) have provided the most authoritative data on forest cover change in the tropics (FAO, 1982a,b, 1993, 2001).[2] For baseline estimates of forest cover and forest cover change in each country, FAO officials have relied on work carried out by forestry officials in each national government. The quality of these country-level estimates has varied substantially across nations, and the unreliability of some estimates has raised questions about the value of the FAO's Forest Resource Assessments (FRA1980, FRA1990, and FRA2000). Only about one third of the countries with large rain forests have actually measured the extent of their forests more than two times since 1970, so only these countries can provide FAO officials with comparable measures of deforestation during earlier and later periods. Under these circumstances, FAO officials have used combinations of less reliable and less valid measures to estimate changes in forest cover. They have used a formula based on rural population density to compute rates of deforestation. They have asked experts to estimate a country's deforestation rate. They have also used the findings from land use surveys conducted in the 1970s to project land use trends in the 1980s and 1990s. The projections from earlier surveys inevitably become less accurate with increases in the period of time since the last survey.

Despite these criticisms of the FAO's measurement procedures, the FRA data still have value. Measurement techniques in individual countries have improved during the past two decades, with some large countries (India and Brazil) now providing reliable estimates of forest cover from remote sensing images (Downton, 1995). The data from the 1980, 1990, and 2000 FRAs also provide the only comprehensive record over time of change in tropical forests. For this reason, the FRA data, despite their limitations, remain indispensable as indicators of historical changes in tropical forest cover.

Comparisons of the FAO's estimates of deforestation rates from the 1980s and the 1990s pose some special challenges because the FAO altered their definition of what constitutes a forest between their 1990 and 2000 surveys. They used a more inclusive definition of dry, open woodlands in their 2000 survey than they did in their 1990 survey (FAO, 2001). This change in definitions led to some dramatic increases in reported forest areas in the eastern and central African countries that contain large amounts of open woodlands. The larger estimates of total forest in turn depressed deforestation rates, because the FAO calculated the annual percentage declines during the 1990s from these larger, redefined forest areas.[3] The redefinitions of forest area had an effect only in some East and central African countries, lowering their estimated deforestation rates by about 0.2 percent per year. The other regional estimates appear unaffected by the change in definitions.

The other sources of inaccuracies in the FAO's estimates, although not random, characterize countries in every region and seem particularistic in their origins, so they produce more noise than bias (Katzer, Cook, and Crouch, 1998:64–69).[4] Because noise reduces the accuracy of all estimates, small differences in the FAO estimates are probably not significant, but large differences probably reflect real differences in tropical landscapes. With this cautionary note in mind, the large, interregional differences in the FRA data, summarized in table 1.1, serve as a starting point for the analysis.[5] These rates of change differ considerably across regions. With the exceptions of south and Southeast Asia, the regions containing the largest blocks of rain forest show the slowest rates of deforestation.[6]

The differences in regional rates of deforestation in the FRA2000 should not be surprising given that reports of deforestation have long emphasized the diverse conditions that drive the phenomenon in different regions. During the 1970s, large landowners carved cattle ranches out of forests in the Amazon basin, and loggers, followed by shifting cultivators, devastated the forests of insular Southeast Asia. In West Africa, migrants to the forest zone planted cacao after clearing out undergrowth and re-

Table 1.1

Annual Rate of Forest-Cover Change in the Tropics by Region, 1980–2000

	Forest Size* (km²) 1990	Deforestation Rate (%) 1980–1990	Deforestation Rate (%) 1990–2000
Central America and the Caribbean	747,240	–1.4	–1.1
Amazon–Orinoco Basin	8,520,310	–0.7	–0.4
West Africa	658,770	–0.8	–1.5
Central Africa	2,275,320	–0.5	–0.3
East Africa	1,784,720	–0.9	–1.1
South Asia	787,800	–0.6	–0.1
Southeast Asia	2,590,620	–1.2	–0.9

SOURCES: For 1980–1990, Food and Agricultural Organization (FAO). 1993. "Forest Resources Assessment, 1990—Tropical Countries." FAO Forestry Paper, #112. Rome; for 1990–2000, accessed at www.fao.org/fo/fra.
    *The figures for forest size include plantations as well as natural forests.

moving some trees from the canopy. The striking differences between the regions in the size and location of their rain forests also contributed to an impression of diverse deforestation processes.

As entrepreneurs who deal in tropical commodities have extended the global reach of their enterprises in recent years, people in different regions have begun to destroy forests in more uniform ways. Asian timber firms, for example, have extended their destructive practices from Southeast Asia to Africa and Latin America during the past 10 years. Fiscal disciplines imposed by the International Monetary Fund (IMF) on indebted governments on all three continents have encouraged these governments to subsidize large plantations and logging operations because these enterprises generate foreign exchange through the export of crops or timber. The same set of fiscal pressures has caused governments on all three continents to reduce the services and programs that they provide for rural populations.

Whereas globalization has caused some convergence in processes of land-cover change, regional differences in economic trajectories have contributed to divergence.[7] During the past two decades, people in Southeast Asia have become much wealthier, whereas Africans have lost ground economically and Latin Americans have struggled to preserve earlier economic gains amid growing income inequality. Other potentially important differences, for example in rural population densities, have persisted between the regions. Taken together, the differences in economic trajectories, population densities, and forest size suggest that deforestation processes continue to vary decisively along regional lines.

Scientific Studies of Tropical Deforestation

If forest cover change seems to vary so much along regional lines, why haven't concerned scientists carried out more regional studies of these processes? Like many other literatures in the social sciences, the literature on tropical deforestation has a bifurcated quality (Ragin, 2000:25). It contains almost 600 case studies of tropical deforestation in particular locales, and more than 400 studies whose conclusions appear to apply throughout the globe. Relatively few studies examine deforestation processes at the regional level. Scientists rarely conduct regional studies, because the methods for conducting them are not well established compared with the methods for carrying out local or global studies. Local researchers typically leave their homes to conduct field research on deforestation in small, exotic places, working among people whom they get to know well. In contrast, global researchers often never leave home at all, crafting intricate models about how deforestation proceeds in far-flung places from data collected by international organizations.

As Katrina Brandon (2000) has pointed out, most field researchers carry around in their heads an image of the process and the problem that derives directly from the places where they have seen forest clearing occur. For these researchers, global forest problems are scaled-up versions of the local processes that they have witnessed. Although grounded in direct observation, these perceptions of larger processes miss important variations between communities in landscape change. Local processes of deforestation recur, with significant variations, in place after place within regions. These regional dimensions become clear only if we aggregate the local studies and synthesize their findings.

The stay-at-home global researchers usually produce studies that include all places but do not seem to apply particularly well to any one place. Many of these studies use data sets on forest cover change within nations to draw conclusions about the forces that drive deforestation across nations. These studies, ably reviewed by Kaimowitz and Angelsen (1998), usually conclude by identifying one set of variables as more important than another set of variables in driving deforestation. Using these techniques, analysts have identified political instability (Deacon, 1994), population growth (Bilsborrow and Okoth-Ogendo, 1992), high levels of international debt (Kahn and McDonald, 1995), and slow rates of urbanization (Erhardt-Martinez, 1998) as important causes for rapid deforestation. Other work, favored by economists in particular, clarifies the economic logic that drives declines in forest cover (Walker, 1987; Ehui, Hertel, and Preckel, 1990; Jones and O'Neill, 1994; Parks, Barbier, and Burgess, 1998). These variable-centered analyses advance our understanding of deforestation processes by identifying and discriminating between theories that do and do not have empirical support.

As this literature has accumulated, one weakness in it has become more salient. Variable-oriented analysis, intent on prioritizing approaches, does a poor job of identifying patterns of conjoint causation.[8] Deforestation is almost always conjunctural in its causation (Geist and Lambin, 2001). In other words, in any given situation, several different causal forces combine to cause land-cover change. Statistical analyses, intent on identifying the most important variables, miss these patterns of conjoint causation, so their findings do not clearly apply to particular situations. Under these circumstances, analysts need an approach and a method that help them identify the different combinations of causal forces that propel land clearing in tropical regions. Middle-range theories developed through comparative historical analyses provide one way to explore and articulate these patterns of conjoint causation.

Middle-Range Theories of Tropical Deforestation

Neither the local nor the global traditions of research on tropical deforestation encourage syntheses. Field researchers express concern for the historical particulars that inhere in the cases that they studied, so they are reluctant to generalize across case studies. Global analysts find it difficult to bridge disciplinary divisions in theory and method. Professionals from each discipline promote their discipline's favorite variable. Together, the local and global traditions have given rise to a fragmented understanding of deforestation processes. This pattern recurs across the "problem-focused" endeavors in the social sciences, and, as Patricia Gober points out, it causes us to miss opportunities for effective policy science. Referring to geography, Gober (2000:1–2) writes, "Fragmentation and compartmentalization come at an extraordinarily inopportune time for our discipline because they limit our ability to respond effectively to the growing demand in science and society for a more synthetic perspective, one that identifies creative linkages between the human and natural systems that govern our world."

Acknowledging the fragmented state of knowledge about tropical deforestation, the International Geosphere-Biosphere Programme convened a meeting of 25 leading researchers in 2000 in an effort to go beyond "case-by-case" data (Lambin et al., 2001). Comparative historical analyses provide a way out of this impasse. Because comparative methods provide a systematic way to aggregate cases and analyze patterns across them, they prevent us from getting lost in a sea of particulars. Because historical analysts can ignore statistical constraints that call for small numbers of variables and large numbers of cases, they find it easier to combine variables and examine patterns of conjoint causation in their analyses of particular cases. The historical focus also directs the analysts' attention toward those trends that over time encourage or discourage the destruction of forests. By highlighting these trends, analysts prepare the ground for new directions in policy making that try to promote conservation by capitalizing on preexisting historical trends in a place.

Combining the comparative and the historical into a comparative historical analysis will not produce a global synthesis of landscape transformation in the tropics, but it will help us identify the middle-range patterns that extend across the different tropical regions. Robert K. Merton and a series of distinguished predecessors have extolled middle-range research as one of the basic building blocks of science. In *Theaetetus,* Plato (cited in Merton, 1968:56–57) observed that "particulars are infinite, and the higher generalities give no sufficient direction; ... the pith of all sciences, which maketh the artsman differ from the inexpert, is in the middle propositions."

John Stuart Mill concurred (Mill, 1865:454–455, cited in Merton, 1968:57): "The lowest generalizations, until explained by and resolved into the middle principles of which they are the consequences, have only the imperfect accuracy of empirical laws; while the most general laws are too general, and include too few circumstances, to give sufficient indication of what happens in individual cases."

Middle-range theories are particularly useful in policy-oriented research, as Bacon noted (Bacon, *Novum Organum,* cited in Merton, 1968:56): "For the lowest axioms differ but slightly from bare experience while the highest and most general are notional and abstract and without solidity. But the middle are the true and solid and living axioms on which depend the affairs and fortunes of men."

George Lewis, another nineteenth-century philosopher (cited in Merton, 1968:57), characterized middle-range theories in the following terms:

"Instead of being mere jejune and hollow generalities, they resemble the Media Axiomata of Bacon, which are generalized expressions of fact, but, nevertheless, are sufficiently near to practice to serve as guides in the business of life."

These axioms, or guides, that result from middle-range research would be what some thinkers refer to as heuristics. Faced with decisions about what type of conservation policy to pursue in a particular region, decision makers should be able to use the trajectories of change outlined here in a heuristic way to craft locally appropriate policies. In other words, middle-range research on change in tropical rain forests should provide regionally specific rules of thumb that policymakers can use as shortcuts in devising policies to conserve tropical rain forests (Jaeger et al., 2001:152–153).

The receptiveness of policymakers in corporations, nongovernmental organizations, and governments to policies derived from middle-range analyses may have a geographic basis, because the boundaries of middle-range investigations often coincide with the geographically defined domains of policymakers. For example, the outline of a regional path of change in south Asian forests should elicit interest among Indian policymakers because India covers most of south Asia. For this reason, studies that outline regional processes should seem particularly relevant to bureaucrats in regional development banks and agencies, to the staffs of international environmental organizations who maintain regional offices, to government officials who follow the initiatives of neighboring countries in forest policy, and to corporate executives who organize their work into regional divisions. Given this coincidence and the large differences between regions in processes of forest destruction and degradation, middle-range research on tropical deforestation has a better chance of influencing conservation policy-making by powerful decision makers than does research conducted at either global or local scales.

2
Theory and Method in Studying Regional Deforestation Processes

The "tropics" are not a plot of convenient forest in Costa Rica; they are an enormous realm of patchiness, and any theoretical thinking based on presumed general properties is bound to become an in-group exercise in short-lived futility.
PAULO VANZOLINI, quoted in A. FORSYTH AND K. MIYATA, *Tropical Nature* (1984:7)

This chapter lays the conceptual groundwork for the subsequent empirical analyses of regional deforestation processes. It defines terms and summarizes the literature on tropical deforestation over the past 20 years. The summary begins with a description of variable-oriented explanations for deforestation, providing brief characterizations of economic, political, and demographic arguments for its causes.

Then I discuss recent attempts to move beyond variable-oriented approaches through the development of agent- and event-oriented approaches that stress the ways in which multiple variables interact to produce forest destruction or regeneration in places (Geist and Lambin, 2001). A description of the event-oriented historical approach used here follows, with special attention to its theoretical foundations in the comparative method and structuration theory. The chapter concludes with a brief discussion of how the qualitative comparative method can, through the application of Boolean algebra, reduce the complexity of the findings in the accumulated case studies to identifiable regional trajectories of forest cover change.

Definitions: Tropical Forests, Deforestation, and Degradation

Tropical Forests focuses on changes in *land cover,* the physical features of landscape, but the explanations for these changes involve analyses of *land use,* the activities that humans undertake to earn livelihoods from the land (McConnell and Moran, 2000). Forest cover varies in type from place to place in the tropics, and all types have experienced increased human pressure over the past 50 years. Accordingly, this study examines a wide variety of human-induced changes in forest cover. It looks at changes in the humid, closed forests that cover large river basins and tropical islands. It also analyzes changes in the tropical, dry forest formations of Central America, south Asia, and East Africa. It examines changes in the submontane and montane forests of tropical Africa, Asia, and Latin America. It also analyzes changes in disturbed and undisturbed forests in regions like Southeast Asia with appreciable numbers of shifting cultivators.

Different observers looking at the same process of forest cover change often arrive at different assessments of the change. Where one person sees ecological restoration, another person may see continuing impoverishment and loss of biodiversity. The different perceptions may stem from the different definitions of deforestation and biological decline used by observers. The Forestry Division of the Food and Agricultural Organization of the United Nations (FAO) has, since the onset of the biodiversity crisis, defined *deforestation* as the permanent reduction in forest cover to a point where trees cover less than 10 percent of the land in a place (FAO, 1993:10). Critics have contended that, by setting such a low threshold for tree cover, this definition neglects the ecological damage that occurs through the partial destruction of stands of trees (Jacobs, 1988:11; Barraclough and Ghimire, 1995:11). Fires, intensive firewood collection, and selective logging in a place extinguish some species and sharply reduce the extent of vegetation in a place, but by the FAO's definition these places remain forested. These definitional disputes disappear if the scope of the study includes forest degradation as well as forest destruction. Fires, firewood collection, logging, and replacement of primary forests with tree plantations all impoverish biological communities in forests and in this sense degrade them. Because degradation entails significant biodiversity losses, it belongs within the purview of this study.

The text uses the FAO definitions for related phenomena. *Afforestation* occurs when people establish tree plantations in places that previously had few or no trees. *Reforestation* occurs when secondary growth replaces portions of a primary forest after logging or a fire (FAO, 2001). Cutover places with regenerating forests never undergo deforestation, because the period with few or no trees is temporary. Some errors will occur with this classificatory scheme. A place that is forested at T1 and cleared of forest at T2 would be considered *deforested*. In most instances, this classification will be correct because the land will remain without trees for a considerable period of time, but in a few instances, deforested areas should probably be considered to be "reforesting" areas.

Theories of Tropical Forest Decline

Soon after alarmed ecologists began reporting sharp declines in tropical forest cover in the 1960s and early 1970s, observers began to identify causal agents. In an early analysis of the problem, Norman Myers identified growing populations of "shifting cultivators" as the primary force behind rapid rates of land clearing in the tropics (Myers, 1979, 1984). Other analysts countered that elite controlled corporations and governments had caused most recent deforestation through large-scale resource exploitation and publicly financed mega-projects (Plumwood and Routley, 1982). Profit-hungry corporations cleared large tracts of rain forest to extract valuable timber and establish plantations to produce crops such as oil palm. Governments intent on establishing control over peripheral places promoted land settlement schemes in remote, sparsely settled, rain forest regions.

As increasing numbers of academic analysts began to address the problem during the 1980s, the language of disciplines and variables began to creep into studies of the deforestation problem. Analysts would typically identify a variable or a small set of variables such as commodity price changes, tenure insecurity, or population growth as the primary forces behind accelerated rates of deforestation. In the more quantitative analyses, authors would then spell out how a particular variable, such as tenure insecurity, affected processes of deforestation across a variety of contexts. The typical forms of these economic, political, or demographic explanations are outlined next.

Economic Explanations for Tropical Deforestation

Over the past 20 years, most analysts have begun to argue that tropical deforestation occurs primarily for economic reasons. Entrepreneurs, heads of households, and individuals make decisions to exploit or destroy forests in response to economic opportunities generated by market trends and institutional factors (Lambin et al., 2001:2). When people choose to convert forests into fields, they in effect make an investment. The short-run return on the investment comes from the sale of crops grown on the land (Angelsen, 1999:207; Wunder, 2000). The conversion of forests into fields may also yield long-run returns if people use the proceeds from crop sales to invest in assets, such as their children's education, which then yield high returns in future decades (Barbier, 1993:2). For these people, the rapid destruction of tropical rain forests represents the economically optimal course of action.

The short-run returns of forest destruction promise to be high only when the cultivator has easy access to markets in central places (von Thunen, 1875), so road building, which eases access to markets, stimulates the conversion of forests to fields in corridors along the new roads. This locational imperative also suggests an eventual end to the land clearing. After a region experiences extensive deforestation, the remnant forests may exist only in rugged, inaccessible mountain regions. The cost of gaining access to these lands and then cultivating them may be so high relative to the expected profits that no one will try to clear them, so forests will persist in these places. Similarly, the value of wood products will increase relative to agricultural commodities as the amount of forest declines in a region. Eventually, the wood products may become so valuable that landowners will decide to preserve patches of forest and plant trees in old fields (Hyde, Amacher, and Magrath, 1996:224).

And when will that happen? After massive numbers of species extinctions have occurred and abrupt climate change has begun? These questions point to a weakness in the economic argument. It assumes that markets have appropriately evaluated the environmental services that tropical rain forests provide (e.g., sheltering potentially valuable species and stabilizing the climate by sequestering carbon in the midst of massive carbon emissions by the industrial world). As a number of analysts have pointed out (Barbier, 1993:4–5), markets tend to underestimate the value of environmental services, and, in so doing, they condone higher levels of forest destruction than they should. These market failures provide a rationale for government intervention to save rain forests.

This critique does not fault economic theorists for misrepresenting the circumstances that cause people to destroy rain forests. Numerous empirical studies attest to the importance of economic motives for changes in forest cover. People clear land to produce wood products for urban markets in Malawi (Abbot and Homewood, 1999); smallholders clear land along roads in the Peruvian Amazon to produce crops for urban markets (Maki, Kalliola, and Vuorinen, 2001), and landowners create woodlots in the least forested countries when forest products become valuable (Rudel, 1998). Clearly, the economic perspective is necessary, but is it also a sufficient explanation for tropical deforestation? In some settings it may be, but in most settings it clearly is not. In the latter instances, causation is conjunctural, dependent on "the combined effect of various conditions, on their intersection in time and space" (Ragin, 1987:25). The intersecting impacts of structural adjustment policies and commodity price trends provide a good example of conjoint causation. Political events during the 1980s, in this case structural adjustment agreements, led governments in Bolivia and Ghana to increase their subsidies to export sectors, soybeans in Bolivia and timber in Ghana, which in turn created economic conditions—favorable price trends—that caused producers to expand the size of their operations at the expense of the forests (Kaimowitz and Smith, 2001; Owusu, 1998).

Political Explanations for Tropical Deforestation

Whereas economic explanations examine market dynamics, political explanations focus on organizations that either directly or indirectly influence decisions to clear land. The jurisdictions for officials in these organizations range from local to transnational arenas. At the local level, village committees regulate access to small plots of forest. At the national level, elites within government ministries shape decisions about forest exploitation (Dauvergne, 1997). At the international level, policymakers in transnational corporations, international nongovernmental organizations, and the International Monetary Fund influence rates of forest destruction.

Given the prevalence of weak states in the tropics (Migdal, 1988), institutional weakness offers an obvious explanation for rapid rates of deforestation.[1] With weak institutions unable to control access to forests, conditions of open access prevail, and people with capital have large advantages in competitions over valuable forest resources. Although laws designate large tracts of forested land as reserves to be exploited in limited ways through government-granted concessions, entrepreneurs circumvent these regulations by constructing patron–client networks in which they pay government officials to give them a free rein in extracting timber (Dauvergne, 1997; McCarthy, 2000). Political turmoil can overturn these arrangements, as they did in the late 1990s in Indonesia, when the prospects for reform after a regime change made the re-imposition of rule by law seem more likely. Alternatively, if political conflict weakens the state still further, as it did in the early 1990s in Cambodia and Nicaragua, the turmoil can strengthen short-term clientilist arrangements, leading to more rapid exploitation of forested lands.

Political conditions shape the pace and location of forest destruction in other ways as well. In frontier settings where the central government finds it difficult to establish law and order, insecurity about ownership induces people to extract and sell valuable wood as fast as they can. The same insecurity has induced people throughout Latin America to clear land and plant crops or pasture in efforts to strengthen their legal claims to land (Southgate, 1998). Because a great deal of deforestation occurs in remote frontier settings in Asia and Latin America, where central authorities exert little control, pervasive tenure insecurity may explain a substantial amount of tropical deforestation (Kaimowitz, Faune, and Mendoza, 1999).

Local as well as national political conditions can affect the status of forests (Gibson, McKean, and Ostrom, 2000:1–26; Ostrom, 1999). Village committees with legitimate authority can restore forests to health by establishing and enforcing rules about forests, but the committees' authority to do so extends only a certain distance from their villages. The location of forests close to or far from village centers can affect the amount of social control that village authorities exert. In the hill country of Nepal, local committees have begun to restore forests close to village centers at the same time that forests situated on ridgelines far above the villages continue to degrade under conditions of open access (Jackson et al., 1998).

Because people often use institutional means to pursue economic ends, political forces often work in tandem with economic conditions to produce forest cover change. For example, the absence of institutional controls in open-access situations only accelerates deforestation when an economic variable such as high prices for wood make the exploitation of a particular tract of forest attractive to entrepreneurs. Although political explanations frequently emphasize the role of local institutions in initiating land-cover change, the geographic scope of these explanations can embrace entire regions. As Thiesenhusen (1991) has pointed out, systems of land tenure and community governance vary in fundamental ways between long-settled, densely populated rural areas in Asia and recently settled, sparsely populated rural areas in Latin America. Long-settled, densely populated rural regions in south and Southeast Asia have elaborate institutions for community governance and few unclaimed resources compared with the more recently settled and sparsely populated communities of colonists in Latin America.[2] These variations in the local institutional presence exert some influence over patterns of forest destruction and regeneration in the two regions.

Demographic Explanations for Deforestation

When Norman Myers first brought the destruction of tropical rain forests to the world's attention during the late 1970s and early 1980s, he identified expanding populations of shifting cultivators as the driving force behind forest destruction (Myers, 1979, 1984). As the number of people grew who cleared small plots of forest (0.25 to 0.5 hectares) each year to cultivate crops, forest cover declined. After 2, 3, or 4 years, smallholders moved on to another plot of land, planting pasture on the old fields or allowing regrowth to occur on the now fallowed lands. When cultivators allowed fields go into fallow, they intended to reuse the fields after resting them for a period of time. With more mouths to feed, smallholders reduced the amount of time that the land lay fallow, so secondary forests did not become fully established before smallholders cleared the land again. Under these circumstances, forests gradually became permanent fields as a growing population filled a once forested landscape with fields.

Population growth also contributed to deforestation by providing the impetus behind long-distance migration to forested frontier regions. Streams of migrants intent on acquiring lands for cash cropping have destroyed large tracts of forest in West Africa (Raison, 1968; Hill, 1963). Similar patterns have also characterized Southeast Asia and Latin America. Government-sponsored colonization schemes designed to settle peripheral forested areas brought large numbers of smallholders to forested regions in the Amazon basin and the outer islands of Indonesia. Once established, the new settlements attracted additional migrants from the same sending areas in rural Java, the Andean valleys, and southern Brazil.

The migrants came from the growing numbers of rural people who, thanks to the introduction of antibiotics and improved medical services, survived to adulthood after World War II (Bilsborrow and Okoth-Ogendo, 1992:42). While the migrants' presence in rural sending areas may be attributable to declines in mortality, they clearly move to the forest frontier in pursuit of economic opportunities, so conjoint causation featuring population increase and economic growth best explains the resulting deforestation in colonization zones. A somewhat similar pattern of conjoint causation may explain continued land clearing in these zones. In these instances, growing populations of urban consumers create demand for agricultural commodities that encourages small farmers to expand land under cultivation at the expense of the remaining patches of forest on their farms (Faminow, 1998).

Again questions of scope arise. When and where do demographic variables have their greatest impact on processes of tropical deforestation? In south Asia and East Africa, where small remnant rain forests persist in densely populated rural landscapes and large numbers of poor rural people eke out existences on the land surrounding the forests, changes in external controls over the forest can quickly lead to forest degradation and destruction. For example, political instability in Uganda in the late 1970s and in Ethiopia in the early 1990s persuaded local people to log forest reserves in Uganda (Struhsaker, 1997) and eliminate reserves in Ethiopia (Kebron and Hedlund, 2000). The potential for this type of exploitation hinges in part on the size of the populations living in close proximity to the forests. For these reasons, analysts continue to point to the presence of large rural populations living in close proximity to the forests when they assess the prospects for forest preservation and conservation in the region (Balmford et al., 2001).

In other places, the direct influence of population growth on deforestation may be declining. Both Latin America and Southeast Asia have experienced dramatic declines in fertility since 1970, so the pool of potential migrants to forest frontiers must be declining because the overall pool of rural youth is declining in size (Sunderlin and Resosudarmo, 1999:158). Although the direct influences of population growth on forest cover change may be weakening, the indirect influences of population growth may be strengthening. Expressed through markets, these impacts occur when increasing demands for agricultural commodities by growing populations of urban consumers spur agricultural expansion. Recent land clearing by small dairy farmers in Honduras (Humphries, 1998) and in Brazil (Faminow, 1998) has followed growth in the size of nearby urban markets.

Until recently, most studies that have argued for a causal connection between population growth and tropical deforestation have not provided detailed, empirical accounts about how population growth has spurred land clearing. This weakness has been addressed in several different ways during the past 10 years. A study of forest cover change in Costa Rica during the 1970s and 1980s demonstrated a positive association between concentrations of landless peasants and the destruction of nearby forests (Rosero-Bixby and Palloni, 1998). Variations in household composition also influence land-clearing rates (Perz, 2001). Clearing primary rain forest from the land is physically taxing work that only young people at the peak of their physical powers will routinely undertake. For this reason, only households and communities that have an abundance of young workers will clear much land. If emigration or declines in birth rates reduce the numbers of young workers in a household, the likelihood of forest clearing on the household's land also declines.

Linking these household dynamics to politically destabilizing events, changes in access to urban markets, and growth in the numbers of consumers should help specify the role of population growth in forest destruction. The preceding examples from Ethiopia, Uganda, Honduras, and Brazil illustrate the conjoint, regionally specific nature of demographically driven causal paths to rain forest destruction. As this book and other recent works (Geist and Lambin, 2001) make clear, most satisfactory explanations for tropical deforestation take this form.

Explaining Regional Processes of Deforestation: Theoretical Underpinnings and Methodological Procedures

Theoretical Points of Departure

The variable-oriented explanations for tropical deforestation compete to some degree with one another. A long series of studies at different levels of aggregation test these theories against one another. The quantitative studies in this analytical tradition assume that "the effect of a cause is the same across different contexts" (Ragin, 1987:166). The context may be a place or a period of time. The authors of these studies calculate the magnitude of the effects that different variables have on processes of deforestation and on the basis of these analyses declare one approach or set of variables superior to another. The consensus paper that emerged from the recent Stockholm workshop on land-cover change does not crunch numbers, but it continues this analytical tradition by declaring economic approaches superior to demographic approaches for understanding the forces that drive tropical deforestation (Lambin et al., 2001).

The value of the variable-oriented approach lies in the high level of generality in its concluding statements. For example, the conclusion in the workshop paper (Lambin et al., 2001) that people clear land in response to economic opportunities provides readers with a rule of thumb to use in their thinking about tropical deforestation. Alternatively, this conclusion could be regarded as so general as to be vacuous. It applies to people in such a wide variety of circumstances that it provides researchers and policymakers with little guidance about critical points for possible intervention in processes of landscape change.

The accumulation of hundreds of studies, each with its own set of variables and combinations of variables that influence forest cover change, often overwhelms analysts and provokes a search for simplicity, usually by refocusing on actor- or situation-specific analyses. Then they engage in a process of progressive contextualization, adding features of the context to the analysis to see how they affect land clearing behavior (Vayda, 1983). For example, after reviewing 150 multivariate analyses of tropical deforestation, Kaimowitz and Angelsen (1998:90–95) chose to summarize the reported effects by focusing on how all of these variables would influence actors' decisions to clear land.

The increase in interest in agent-based models (ABM) of land use and land-cover change probably stems from a similar impulse (Parker, Berger, and Manson, 2002; Parker et al., 2003). These analytical strategies have the virtue of focusing analysts' attention on one actor's decision making rather than diffusing their attention across myriad competing variables and situations. These modeling efforts have had the additional benefit of producing simulations that forecast land use changes as much as 25 years into the future (Vosti, Witcover, and Carpentier, 2002). Unfortunately, the simplifications required in agent-based models and frameworks can come at a high empirical price. To date, much of the agent-based modeling has focused on a prototypical shifting cultivator (Walker, 1987, 2003). As Geist and Lambin (2001) point out, analysts should probably shift their attention to networks of actors, some of whom occupy positions of influence in the political coalitions that secure commitments to build penetration roads into rain forest regions. The challenges of integrating this level of complexity into agent-based models are formidable (Coucletis, 2002).

Helmut Geist and Eric Lambin (2001) offer an alternative approach to modeling deforestation processes that closely resembles the approach used here. Like the agent-based modelers, Geist and Lambin seek simplicity by focusing on specifics, but unlike ABM analysts, they focus on situations, not on actors. In effect, they practice what Vayda refers to as "event ecology" (Vayda and Walters, 1999). They try to explain ecologically significant events—in this instance, the destruction of tropical forests. Working outward from these instances, they too engage in progressive contextualization, identifying first the proximate causes and then the underlying causes in a meta-analysis of 152 case studies of tropical deforestation. They identify proximate causes such as wood extraction, road building, or agricultural expansion, and underlying causes of a more disciplinary nature such as economic trends or demographic pressures.

Several features of this analytical approach recommend it for use. The focus on proximate causes grounds the study in specific instances of forest destruction at the same time that the meta-analysis increases the range of instances under study. In this sense, the Geist and Lambin approach promises to deliver both breadth and depth in the analysis of tropical deforestation. The focus on conjoint causation with multiple, interacting causes in specific places takes us away from the reified competition between variables and disciplines that inheres is so much academic research and frequently gives it an air of unreality. Finally, the emphasis on proximate causes seems salutary because it encourages observers to wrestle with the complexities of particular cases in developing their interpretive schemes.

Tropical Forests exhibits many of the features of the Geist and Lambin study as well as some significant differences. Like the Geist and Lambin study, it carries out a meta-analysis of case studies. Like that work, it is an example of "event ecology," but it includes case studies of forest regeneration as well as forest destruction. It adds a historical dimension to the meta-analysis, distinguishing between forest cover changes occurring during the 1970s and 1980s and those occurring during the 1990s.

Whereas Geist and Lambin (2001) report regional differences in a wide-ranging discussion of the different forces that drive deforestation, *Tropical Forests* defines the regions differently and focuses the analysis on the different regional combinations of driving forces. Finally, although both studies embrace the idea of conjoint causation, *Tropical Forests* uses a different method, the qualitative comparative method, to identify the particular combinations of forces that drive forest cover change in places.[3]

In other words, *Tropical Forests* presents a comparative historical analysis of recent forest cover changes in the tropics. It asks, Where and when do particular explanations for deforestation apply? Rather than ignoring variations between geographic and historical contexts in studying how causes work, this approach tries to identify differences between contexts in the configuration of causes that explain tropical deforestation. Rather than declaring the economic approach superior to the demographic approach, a comparative historical approach to the problem will try to identify the historical and geographic contexts in which economic changes rather than population growth shape forest cover trends.

Deforestation processes certainly vary across regions, and they also vary through time. Structuration theory (Giddens, 1984) provides a set of useful conceptual tools for analyzing the temporal dimension of deforestation processes. Analysts acknowledge the historical dimension in tropical deforestation when they talk about the "path-dependent" nature of land-cover change (Lambin et al., 2001; Rudel and Roper, 1997a). Humans "make history, but not in circumstances of their own choosing" (Marx, 1852, quoted in Giddens, 1984:xxi). Driving forces shape land use changes, but they do so through the active involvement and understanding of peasants, corporate employees, and government officials who live and work in rain forest regions. The "daily paths" that individuals follow in earning their livelihoods cumulate into "life paths" that from generation to generation reproduce or alter important individual and institutional practices such as agricultural expansion or extractive logging (Giddens, 1984:367). The configurations of causal forces behind these practices are "not predetermined, but are formed in a path-dependent way as each actor, with more or fewer resources at his or her command, shapes a new social structure by drawing on the simultaneously enabling and constraining hand of the old" (Molotch, Freudenburg, and Paulsen, 2000:791).

For example, as loggers deplete the commercially valuable wood near roads in northern Sumatra, they alter the economic choices available in future years to their descendants (McCarthy, 2000). When middle-aged peasants in Honduras choose to sell recently cleared land and move farther into the forest to start another farm (Sunderlin and Rodriguez, 1996), they repeat a pattern of activity that they learned at an earlier age. In these ways, households and corporations chart a particular path or trajectory of economic exploitation for each place. At each step in the path, actors contend with "sunk costs" that shape how they decide to use forested land (Barham, Chavas, and Coomes, 1998). Actors follow these daily paths through places whose contextual features recur to varying degrees throughout a region. When the individual paths and surrounding contextual features recur to a great degree, a regional process of deforestation becomes visible.

Analysts have long recognized that the location of forests makes a difference in how processes of tropical deforestation proceed, but the people who have designed studies to explore these differences have opted for a simple continental determinism, a study site on each continent. The United Nations Research Institute for Social Development (UNRISD) has carried out case studies of deforestation in Africa, Asia, and Latin America (Barraclough and Ghimire, 1995, 2000). Similarly, the Center for International Forestry Research (CIFOR), as a matter of policy, conducts empirical research for its projects on three continents, Africa, Asia, and Latin America. Although the idea of carrying out field research in diverse settings certainly is commendable, the UNRISD and CIFOR procedures seem to assume that deforestation processes vary primarily along continental lines. Geist and Lambin (2001), in their meta-analysis of research on tropical forest cover change, also assume a kind of continental determinism, breaking their studies down into African, Asian, and Latin American subsamples. This assumption is open to empirical question. Given the tremendous variation in the physical settings of tropical forests in Africa, deforestation processes may, for example, vary more between East and central Africa than they do between East Africa and south Asia. For this reason, an empirical method for distinguishing between distinct regional deforestation processes would be most useful. The method outlined later makes these distinctions.

If the variable-oriented strategies verge on making vacuous generalizations, case-oriented comparative methods can become mired in the specificities of particular cases. Researchers investigating individual cases can build complex models of conjoint causation that account for deforestation in the case under study, but the application of these models to other cases remains open to question. To get beyond the individual case studies and realize the analytical potential of a comparative historical approach, it must be combined with a method for arraying and analyzing data from multiple case studies.

A Method for the Meta-analysis

These circumstances call for a meta-analysis, but what kind? Traditional methods for meta-analyses, developed most completely in psychology, cannot be used. To do a conventional meta-analysis, psychologists take the data from the studies under examination and pool them, creating a large sample of respondents. Then they analyze the pooled data. The data requirements for this type of meta-analysis are quite stringent. To pool data from different studies, investigators must have collected data on the same variables, using similar instruments, in somewhat comparable settings. In the nonexperimental social sciences, investigators may focus on the same topic, but both the variables and the measures differ from study to study, so meta-analysts cannot pool the data. For example, students of tropical deforestation have used a variety of methods, including household surveys, key informant interviewing, field observations, and satellite imagery, to collect primary data on tropical deforestation. The variables differ along with the methods, so it is impossible to pool these data and analyze them in a meaningful way.

Rather than abandoning all hope of using formal methods to carry out a meta-analysis, analysts working with more disordered literatures could carry out model-centered rather than data-centered meta-analyses. In this procedure, we pool the models that analysts have developed from their data rather than the data itself. Then we look for systematic patterns across the models. For example, some researchers argue that population growth drives tropical deforestation in their field site; other investigators do not make this assertion. Each study becomes a case and each assertion about the driving forces of tropical deforestation in that case becomes an observation in the meta-analyst's data set. Taken together, the observations in each case present a model of how deforestation proceeds. Comparisons across the cases establish geographic and historical patterns in the models that presumably represent real differences in the processes under study. In other words a comparison of South American models from case studies conducted during the 1980s and case studies conducted during the 1990s might yield insights into how South American deforestation processes changed from the 1980s to the 1990s.

This approach to meta-analyses presumes that field researchers faithfully translate their empirical observations into models. It downplays the importance of discursive tendencies in the literature, arguing in effect that, although fashions in explaining forest cover change certainly shift from decade to decade, the changes in fashion usually have empirical bases. For example, demand for fuelwood became a popular explanation for deforestation in Africa during the early 1980s on the basis of reports indicating that the growing populations of rural people would cut down nearby forests to meet their fuelwood needs. The assumption that forests would suffer at the expense of fuelwood collectors proved to be inaccurate when a series of studies showed that most people collected fuelwood not from forests but from brushy areas near villages. In the 1990s, fewer analysts cited fuelwood as a major cause of forest decline (Arnold et al., 2003). Similarly, analysts stopped talking about new land settlement schemes as causes for forest decline when most governments either abolished or downsized these programs in the late 1980s.

Another problem with discursive origins involves disciplinary differences among field researchers. Economists see markets everywhere and cite them as the cause of changes in forest cover; demographers routinely identify population change as a cause for changes in forest cover. A shift in the relative proportions of economists and demographers studying tropical deforestation from the 1980s to the 1990s could lead to a corresponding change in the relative frequency with which people identify markets or population changes as the forces behind changes in land cover. One partial way around this problem is to focus the comparisons in a meta-analysis on proximate rather than ultimate causes—in other words, to focus on those on-ground activities such as road building or logging that disturb or destroy forests. The disciplinary identifiers of these activities are less obvious, and they are likely to be reported by observers irrespective of their disciplinary background, so proximate causes are less likely to be favored or ignored in a systematic way by analysts from different disciplines. Geist and Lambin (2001) emphasized the identification of proximate causes in their meta-analysis, and they found little evidence of disciplinary biases in their study. *Tropical Forests* follows Geist and Lambin in emphasizing proximate causes, so the likelihood of disciplinary biases would appear to be small. These examples and strategies notwithstanding, there are clearly discursive tendencies in the literature, and, to some small extent, analyses like the ones presented here pick up these tendencies rather than actual changes in the forces driving changes in forest cover.

This approach also assumes that, out of the very different reporting styles of investigators (e.g., some use Greek symbols and others only use words), a coder can create models that are truly comparable. Although these difficulties seem real, the model-centered approach to meta-analyses seems, for all its faults, to offer the only feasible and still systematic way to include hundreds of disparate studies in a meta-analysis. This type of meta-analysis should provide a middle way between the generalities of variable-oriented analyses and the specificities of ethnographic case studies, making it possible to identify regional paths of forest cover change.

The simplest way to compare the models from the assembled case studies is by grouping them into sets of studies that find similar causal configurations for forest cover change. Qualitative comparative analysis (QCA) lends itself to this analytical task for several reasons (Ragin, 1987). It uses algorithms drawn from Boolean algebra to sort cases into minimized sets of factors that in different combinations cause a particular condition. It works best for data sets that range from 6 to 70 cases, the approximate size of the literatures on tropical deforestation in each region. Like our model-centered meta-analysis, it relies on a data set of binary observations.[4] In our study, the binary observation would indicate the presence or absence of a particular driving force such as population growth in a case study of forest cover change. The case would also have a binary observation indicating the presence or absence of tropical deforestation in this instance. To create a data set for a QCA, analysts have to read and code each study for the presence or absence of a long list of potential causes of tropical deforestation. They use these data to construct truth tables that list each study, its location, and the driving forces behind deforestation in that locale. Table 2.1 presents a simplified version of a QCA truth table. It locates case studies in both space and time, in this instance in Ecuador between 1980 and 2000.[5] It also codes the studies for the presence or absence of particular driving forces. Because conjunctural causation characterizes tropical deforestation, most case studies identify a combination of causal conditions.

The truth tables that I have constructed for the analyses of regional deforestation processes in subsequent chapters are much larger than the truth table in table 2.1. The regional truth tables contain information about 20 different driving forces in addition to information about the

Table 2.1

A Simplified Truth Table for Studies of Tropical Forest Cover
Change in Ecuador during 1980s and 1990s

Author	Data from	Driving Forces					Forest Loss
		Small Ag.	Mineral Exploit.	Logging	Road Building	Colon. Programs	
Echevarria, 1994	1980s	No	Yes	No	No	No	No
Rudel et al., 1993	1980s	Yes	No	No	Yes	Yes	Yes
Pichon et al., 2001	1980s	Yes	Yes	No	Yes	Yes	Yes
Hiroaka, 1980	1970s	Yes	Yes	No	No	Yes	Yes
Becker, 2000	1990s	Yes	No	No	No	No	Yes
Sierra, 2000	1990s	Yes	Yes	Yes	Yes	No	Yes
Wunder, 2000	1990s	Yes	No	Yes	Yes	No	Yes
Rudel/Bates, 2002	1990s	Yes	No	Yes	No	No	No

date of a study, its location, and the methods used to collect data. The truth tables include only those studies that analyze some sort of primary data. Studies that rely exclusively on secondary data are not included in the meta-analyses. For example, essays that synthesize the empirical work of others would not be included in the truth tables. For similar reasons, quantitative, cross-national analyses of data collected by the FAO are not included in the truth tables. The tables include studies in which the authors have conducted field research or carried out remote sensing analyses of deforestation at any level of aggregation up to and including the nation. At the end of 2003, the literature contained more than 270 such studies.[6] These studies provide the empirical bases for the descriptions and explanations of forest cover changes in this book. Readers may want to inquire if I included a particular case study in the analysis and, if so, how I coded the causal forces from that study. Readers can answer these questions by consulting the appendix. It presents a list of the case studies included in each regional meta-analysis. It also lists the case studies that fit each causal configuration in each regional QCA.

Boolean algebraic techniques provide a means for uncovering and simplifying the regional patterns of causation across these studies. In some instances, several combinations of causal conditions can be logically subsumed, using the rules of Boolean algebra, under a simpler set of conditions. Through this type of transformation, QCA minimizes the set of causal conditions associated with a particular outcome of interest. For example, table 2.2 presents the results from a Boolean analysis of the truth table for forest cover change in Ecuador between 1980 and 2000. It compares the four studies from the 1980s in table 2.1, three of which reported extensive deforestation and one that did not. It concludes, as reported in part A of table 2.2, that a causal conjuncture of small-scale agriculture, coupled with road building and colonization programs, produced deforestation. Contradictory results characterize another driving force—mineral exploitation. During the 1980s it was present in some, but not all, instances of deforestation. Because the findings for this driving force are inconsistent across the studies, QCA drops it from the minimized expression. This pattern suggests that it is the road building associated with oil fields rather than the oil fields themselves that causes most forest destruction.

Part B of table 2.2 presents the minimized results for the 1990s. It summarizes the causal conjunctures behind deforestation in two expressions. The second expression cites only one factor, small-scale agriculture as a driving force. A further investigation of this particular case would indicate that another factor, insecure land tenure, played a role in this instance of deforestation. This finding would suggest that adding a land tenure variable to the analysis might be analytically fruitful. Comparing the results from the 1970s and 1980s (part A) and the 1990s (part B) indicates some substantive changes in deforestation processes between the two periods. In particular, colonization programs disappeared as a significant driver of deforestation during the 1990s. The number in parentheses at the end of each causal configuration indicates the number of case studies that exhibited that combination of causes.

Tables 2.1 and 2.2 demonstrate how QCA can aggregate causal arguments across case studies and then use Boolean algebra to reduce or simplify the summary expression of these arguments. The findings in part A of table 2.2 are somewhat anomalous in one respect. Typically, a single model of conjunctural causation (e.g., SMALLAG logging ROADS COLON) does not adequately describe the causal forces behind deforestation in any given region. Rather, there is a series of models that together depict all of the processes of deforestation described in case studies from a region. In this sense, QCA is a *diversity-oriented research protocol,* in that it does not suppress unique combinations of factors by focusing on the central tendency, the most common set of causal factors, in the data

Table 2.2
Tropical Deforestation in Ecuador, 1980–2000:
A Qualitative Comparative Analysis

A: Deforestation During the 1970s and 1980s (4/4)
SMALLAG logging ROADS COLON (3)

B: Deforestation During the 1990 (4/4)
SMALLAG LOGGING ROADS colon (3) +
SMALLAG mineral logging roads colon (1)

The QCA is based on Table 2.1.

The "+" in Boolean algebra, and in this table, indicates that the condition in question (deforestation) occurs whenever any of the listed combinations of causal conditions exists.

Upper case indicates the presence of a factor, and lower case indicates its absence. The ratio in parentheses indicates the proportion of studies for that period whose findings agree with the Boolean expressions in the line(s) below it. The number in parentheses after each expression indicates the number of studies whose findings were consistent with the configuration of causes in that expression.

Factors: COLON = colonization program; LOGGING = logging in region; MINERAL = mineral extraction (oil) in region; ROADS = penetration road construction; SMALLAG = smallholder agriculture.

set. In this way, QCA tries to preserve a maximum amount of diversity across cases even as it attempts to reduce the causal patterns in the cases to a simpler set of expressions.[7]

These goals conflict to some degree and an analyst must make choices. A simple set of causal combinations clarifies the conditions under which forest cover change occurs, but, to simplify the set, the analyst may have to exclude the cases that do not conform to the causal conditions specified in the set. The analyst achieves simplicity at the expense of coverage. In this book, the reader can gauge the extent of the tradeoffs in a particular analysis by looking at the proportion of all cases explained by the causal configurations after the title of a QCA. If the proportion indicates a significant number of omitted cases, then I had to omit a number of cases to get a simple expression. In some QCAs, no cases were omitted. For example, the causal configuration in part A of table 2.2 includes all four of the 1980s case studies, so the reported fraction is 4/4.

The ability of Boolean algebra to minimize the findings from a particular set of studies depends on the empirical patterns in the studies. If the set contains studies that report quite different findings from place to place, then Boolean algebra will not be able to simplify the causal patterns very well and a long series of different causal conjunctures will describe deforestation in a region; in this case, one could reasonably argue that the region does not contain a distinctive pattern of deforestation.

Under these circumstances, it makes sense to alter the boundaries of a region, excluding some studies and including others, in an attempt to see if a Boolean analysis of a somewhat different geographic set of studies will produce a more coherent pattern of forest cover change.

These analyses produced two notable departures from the common understanding of geographic regions in the tropics. Processes of exploitation in the coastal forests of Colombia and Ecuador seemed more closely related to Central American processes than to processes that prevail in the landlocked forests of the Amazon basin, so studies from the Colombian and Ecuadorian coastal regions were included in the Central American analysis. Similarly, land use patterns in the more densely settled, coastal locations of Cameroon seemed to resemble the patterns of exploitation in West Africa's coastal forest regions more than the patterns of exploitation in the sparsely populated, often inaccessible forests of central Africa. For this reason, the studies from these regions of Cameroon were included in the West African analysis. This iterative process of analysis outlined the boundaries of the seven regional patterns of deforestation. Table 2.3 lists the regions and their constituent countries.

The existence of seven different regional patterns became apparent in the mid-1990s through an analysis of all of the then-published case studies of tropical deforestation. An analysis of 115 case studies revealed seven distinctive regional trajectories of forest cover change (Rudel and Roper, 1996). The regions were as follows: South America, Central America, West Africa, central Africa, East Africa, south Asia, and Southeast Asia. The following analysis builds on this work. It analyzes the seven regional patterns with data from a now much larger number of case studies.

The population of case studies for any region will contain case studies done at different scales of analysis. For example, the studies of Central America include a country-level, remote sensing study of forest cover change in Costa Rica (Sader and Joyce, 1988) as well as a study of forest cover change in a single valley in northern Honduras (Humphries, 1998). Combining studies done at different scales in a single analysis may sometimes suppress substantively significant variation in driving forces because, as geographers frequently note, analysts see different things when they conduct analyses at different scales.[8] To eliminate this artifactual element from the analyses, one could separate the studies within a region by scale and carry out separate QCAs on each subgroup of studies. Unfortunately, the number of studies carried out at higher scales was so small that this course of action did not seem feasible, so the

Table 2.3
The Countries in Each Rain Forest Region

Central America and the Caribbean	Bahamas	Haiti
	Barbados	Jamaica
	Belize	Honduras
	Colombia (partial)	Mexico
	Costa Rica	Nicaragua
	Cuba	Puerto Rico
	Dominica	St. Lucia
	Dominican Republic	St. Vincent
	Ecuador (partial)	Trinidad and Tobago and Nine
	El Salvador	European dependencies in the
	Grenada	Caribbean
	Guatemala	
Amazon Basin	Bolivia	Guyana
	Brazil	Paraguay
	Colombia (partial)	Peru
	Ecuador (partial)	Suriname
	French Guiana	Venezuela
West Africa	Benin	Guinea-Bissau
	Cameroon (partial)	Liberia
	Côte D'Ivoire	Nigeria
	Equatorial Guinea	Sao Tome and Principe
	Gambia	Senegal
	Guinea	Sierra Leone
		Togo
Central Africa	Angola	Congo (Brazzaville)
	Cameroon (partial)	Congo (DRC)
	Central African Republic	Gabon
East Africa	Botswana	Mozambique
	Burundi	Rwanda
	Ethiopia	Seychelles
	Kenya	Sudan
	Madagascar	Tanzania
	Malawi	Zambia
	Mauritius	Zimbabwe
South Asia	Bangladesh	Nepal
	India	Sri Lanka
Southeast Asia	Brunei	Myanmar
	Cambodia	Papua New Guinea
	Indonesia	Solomon Islands
	Laos	Thailand
	Malaysia	Vietnam

different scales of the case studies remains a potential source of error in the analysis.

These methodological procedures give the analysis breadth but little depth. They make it possible to identify the broad patterns of deforestation that prevail across a region, but they do not provide a detailed understanding of how regional circumstances generate deforestation. To achieve depth in the analysis, I look in more detail at well-documented case studies from the region whose findings represent the larger patterns uncovered in the QCA analysis. This mix of methods will hopefully achieve both breadth and depth in the analyses of regional deforestation processes. The following chapters present analyses of deforestation in each of the seven regions.

3

Central America and the Caribbean: Island and Isthmus Deforestation

We crossed the current to the quiet shore inside a long bend, slipping through the glossy, black water with only the rhythmic push and suck of the old man's paddle blade to mar the silence of the gliding. We were near the river mouth and at first the shores were low plains of grass and ragged mangroves, but, as we moved upstream, the manaca and huiscoyol palms came in, and then real fringe forests, with big trees rising or leaning at the water's edge and hung with liana ropes and swathed in the killing green velvet of vine drapery. At first slowly, then with a rush, the feel of a Caribbean river came back to me.
ARCHIE CARR, *The Windward Road* (1955:58–59)

The Physical Setting of the Forests

Forests once stretched eastward from high peaks in the interior to beaches along the windward shores of the Caribbean Ocean. The prevailing trade winds still bring rain from humid ocean air masses to the coasts and mountains, but the rains nurture much smaller forests than they did 500 years ago. Forests persist in mountain redoubts on Caribbean islands, and a strip of fragmented, moist forest runs north, from Panama to Mexico, along the Caribbean coast of the Central American isthmus. Just to the west, paralleling the moist forests, patches of dry forests and upland pine forests extend from Costa Rica to northern Mexico. Compared with the large rain forests that blanket the Amazon basin, central Africa, and insular Southeast Asia, the tropical rain forests of Central America and the Caribbean are small. Although the largest block of rain forest in Central America runs for more than 1,500 miles from north to south, it extends only 50 to 200 miles inland from the coast. The remnant forests on the Caribbean islands are often mere "spots" of forest cover in rugged mountain topographies. Figure 3.1 charts the reduced extent of the forests.[1]

A History of the Forests to 1980

Humans cleared extensive tracts of Central American forests for the first time about 1,000 B.C. when Mayans began to expand the scale of their agricultural activities. The network of Mayan cities and agricultural hinterlands gradually spread over much of the rain forest region in present-day Honduras, Guatemala, Belize, and the Yucatan peninsula of Mexico. After the collapse of Mayan society about 1,000 A.D., forests reclaimed the old Mayan fields (Turner et al., 2001).

Large-scale human disturbances began again in the sixteenth century, when the rains, warm climate, and proximity of forested land to the sea attracted European adventurers and entrepreneurs to the West Indies. The entrepreneurs subjugated the indigenous peoples and converted coastal forests into sugar plantations that fed the newfound European appetite for sugar. After disease decimated the local Amerindian populations, planters imported slaves from Africa to work on the plantations (Mintz, 1986). On many islands, the landscape became "a mountain range surrounded by sugarcane" (Tugwell, 1946:32). In the nineteenth and early twentieth centuries, descendants of the first settlers moved inland, establishing coffee groves in upland regions of the isthmus and the islands. Land clearing followed long cycles in international trade, waxing when the prices of international commodities such as sugar rose and waning when prices declined. In Jamaica, for example, planters converted large amounts of forested land into sugar plantations during the eighteenth century, leaving few, if any, trees standing in many locales. The forests regenerated in many places when sugar prices fell during the late nineteenth century and planters abandoned their lands (Eyre, 1987:340).

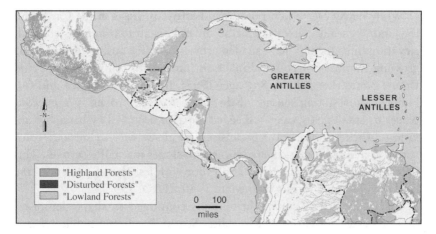

Figure 3.1   The forests of Central America and the Caribbean.

Planters intent on export-oriented agricultural expansion continued to gnaw away at the forests during the early twentieth century, but their identities and their crops changed. New England merchants discovered that they could deliver fresh Central American fruit to urban markets in the eastern United States and Europe. Within three decades, American entrepreneurs had organized a large corporation, the United Fruit Company, and converted big tracts of forest along the humid, windward shores of Central America into banana plantations (Kepner and Soothill, 1935). Extensive tracts of land containing a single variety of banana (Gros Michel) invited invasions by pests, and, by the early 1920s, a root mold, christened Panama disease for its place of origin, had begun to wipe out banana plantations on the Caribbean shores of Central America. Unable to fight Panama disease, companies chose to flee it. First, they established new plantations in hitherto unexploited regions along the shores of the Caribbean. Later, they carved irrigated banana plantations out of the dry forests along the Pacific coast of Central America. In this manner, banana producers created during the first half of the twentieth century a series of banana frontiers, each of which destroyed old growth tropical forests (Tucker, 2000:147–178; Wunder, 2001). With the introduction of the new, disease-resistant Giant Cavendish variety and the increased use of irrigation, pesticides, and fertilizers, growers began to abandon their footloose ways after 1960, and some coastal areas became semipermanent sites for cultivating bananas.

While newly created corporations destroyed forests in the Caribbean lowlands, peasant cultivators farther inland began to thin forested, mid-elevation hillsides and plant shaded groves of coffee bushes to supply the growing numbers of coffee drinkers in nineteenth-century Europe and United States (Bergad, 1978; Paige, 1997). In Central America and Colombia, the resulting mosaic of disturbed forests, broken in numerous places by small fields, easily exceeded the area occupied by banana plantations (Tucker, 2000:181).[2] Because coffee grows well on hillsides, cultivators converted forest-covered mountain slopes into coffee groves. This practice left countries such as El Salvador with less than 5 percent of the land area in intact forests by the late 1940s (FAO, 1948).

Externally driven commodity booms continued to shape deforestation processes in Central America after World War II, but public officials began to play a more active role in the process. Assisted by publicly financed lines of credit and improvements in roads, wealthy families acquired land, evicted smallholders, and, through the sweat of poor migrant laborers from the highlands, converted large tracts of lowland dry forest into cotton fields. By the 1960s, Central American growers ranked third in the world (behind American and Egyptian growers) in the volume of cotton that they exported (Williams, 1986:14, 20–27).

Cattle ranchers converted forests in the highlands and along both coastal plains into pastures during the same period. Again, events in foreign markets triggered the change in land use. Increases in demand for beef, in particular by rapidly expanding chains of fast-food restaurants in the United States, provided a new export market for Central American farmers and the primary impetus behind the expansion in pastures. New refrigeration technologies made it easier to get beef from the pastures of Central America to tables in the restaurants of North America. Public officials strengthened the "hamburger connection" between affluent American consumers and cattle ranchers in Central America by building roads from seaports to the new cattle producing regions (Tucker, 2000:322–332; Myers, 1981). Officers at international development banks provided governments with loans to finance the construction of the new roads. The beneficiaries of the states' largesse were either landed elites or urban professionals with a desire to enter agriculture.

Other public programs promoted cattle ranching among poorer segments of the rural population. Apprehensive about the possibility that the Cuban revolution would inspire similar events in neighboring countries during the 1960s, American officials expanded the United States' bilateral aid programs to Central America and allocated the additional funds to colonization programs. These new land settlement schemes funded the construction of penetration roads into hitherto remote rain forest regions and in some instances helped landless peasants convert tracts of rain forest into small cattle ranches. In this manner, politicians hoped to co-opt the rural poor and, in so doing, prevent a revolution. In the 1960s and 1970s, Guatemala, Honduras, Mexico, and Nicaragua all established colonization programs to open up rain forest regions in the Caribbean lowlands for settlement and deforestation (Schwartz, 1987; Jones, 1989:59–66; O'Brien, 1998).

The hamburger connection is only the most recent link between consumers in the north and producers bent on extracting wealth from forested lands bordering the Caribbean. A historical cascade of export-oriented commodity booms—sugar in the eighteenth century, coffee in the nineteenth century, bananas in the early twentieth century, and cotton and cattle more recently—produced extensive and, at times, rapid deforestation. By the late twentieth century, few countries had extensive tracts of primary forests. Some countries in the region, such as El Salvador and Haiti, had virtually no old growth forests. Countries with larger forests, such as Costa Rica and Puerto Rico with 38 percent and 37 percent, respectively, of their land in forests, had more secondary forests than primary forests (FAO, 2001). The extensive secondary forests and scrub growth provide mute testimony of past export booms that went bust.

Regional Deforestation Processes Between 1980 and 2000

The Retreat of National Governments from Rural Areas

A Boolean analysis of 62 case studies of deforestation in the region, presented in table 3.1, highlights more recent changes in the region's forest cover. More than 80 percent of the case studies focus on mainland places, with particularly large numbers of studies conducted in Mexico (17), Costa Rica (9), and Honduras (8). There were no case studies of land use change in Cuba. Part A of table 3.1 summarizes the findings from 43 studies of deforestation during the 1970s and 1980s, and part B summarizes the findings from 19 case studies during the 1990s. Differences between parts A and B suggest some changes in the forces that have induced people to destroy a region's forests. In particular, the presence of colonization programs (COLON) in the 1970–1980s models and

Table 3.1
*Deforestation in Central America and the Caribbean:
A Qualitative Comparative Analysis*

A. 1970s and 1980s (37/43)
COLON RANCHING SMALLAG (10) +
COLON RANCHING LOGGING (7) +
RANCHING SMALLAG POV (3) +
COLON LOGGING pov (4) +
SMALLAG POV colon (2) +
RANCHING LOGGING smallag pov (3) +
logging ranching colon POV (3) +
RANCHING colon smallag pov (4)

B. 1990s (19/19)
logging SMALLAG pov TENURE (2) +
SMALLAG pov TENURE RANCHING ROADS (3) +
logging SMALLAG pov ranching ROADS (2) +
logging POV TENURE ranching roads (2) +
logging SMALLAG TENURE ranching roads (1) +
logging SMALLAG POV RANCHING roads tenure (2) +
LOGGING SMALLAG tenure ranching roads (1) +
LOGGING SMALLAG pov tenure roads (3)

The "+" in this table indicates that deforestation occurs whenever any of the listed combinations of causal conditions exists.

Upper case indicates the presence of a factor; lower case indicates its absence. The ratio in parentheses indicates the proportion of studies for that period whose findings agree with the Boolean expressions in the lines below it. The number in parentheses after each expression indicates the number of studies whose findings were consistent with the configuration of causes in that expression.

Factors: COLON = agricultural colonization of new regions with government assistance; LOGGING = timber exploitation; POV = absence of nonfarm economic opportunities; RANCHING = cattle ranching; ROADS = penetration road construction; SMALLAG = smallholder agriculture; TENURE = insecure claims to land.

their absence in the 1990s models signals a reduction in state support for rain forest destruction. The four Central American governments that launched state-supported new land settlement schemes in the Caribbean lowlands during the 1960s and 1970s either ended or reduced the scale of these programs during the 1980s. At the same time, governments reduced the number of penetration roads that they constructed into rain forest regions. While decisions to reduce the involvement of the state in new land settlement schemes may have stemmed from increased fiscal pressures on heavily indebted governments in the 1980s, the reduction in state activity may also have resulted from declines in the size of sparsely populated rain forest regions. When forests declined in size, politicians no longer had empty spaces to fill with poor people in politically expedient ways, so they eliminated colonization programs.[3]

The turbulent political conditions in poor countries such as Haiti or Nicaragua (Monaghan, 2000; Kaimowitz, Faune, and Mendoza, 1999) made land tenure precarious, and in the fiscally straitened circumstances of the 1990s, government officials did not improve land tenure security. These conditions explain the salience of tenure insecurity (TENURE in table 3.1) in the QCA for the 1990s. Uncertainties about ownership diminished incentives to intensify production through land improvements because of fears that investors could lose their land and the improvements on it. Under these circumstances, short-term plans to mine a remote area for its timber or deplete new fields of their fertility seemed more attractive than afforestation efforts that yielded a return only if the owner could retain the land for more than a decade. For example, efforts in Haiti to increase supplies of fuelwood through afforestation foundered in part because entrepreneurs continued to supply charcoal to markets in Port-au-Prince by extracting wood from a small number of distant rain forests (Catanese and Perlack, 1993).

Contrasting Trajectories of Change Within the Region

The expansion of global markets contributed to a common environmental history in the Caribbean region, from expanding sugar plantations in the seventeenth century to spreading coffee groves and banana plantations in the early twentieth century. Despite these similarities in market forces, the historical course of forest cover change has often differed dramatically from country to country. Puerto Rico and the Dominican Republic exemplify this tendency. At the beginning of the twentieth century, stark contrasts characterized the landscapes of the two countries. An eroded, treeless, almost entirely cultivated landscape covered Puerto Rico, whereas, 70 miles to the west across the Mona Passage, the "luxuriant" vegetation of an old growth tropical rain forest covered more than 75 percent of the Dominican Republic (Durland, 1922:206). Over the next 90 years, the islands' landscapes evolved in opposite directions. Forests spread throughout the hilly interior of Puerto Rico, covering almost 40 percent of the island's landscape in green, after cultivators abandoned their coffee groves and took nonfarm jobs in urban places. During the same period, old-growth forests disappeared from many places in the

Dominican Republic as small farmers expanded their cultivation of crops such as coffee and cacao. In recent years, as Dominicans moved to cities and the United States, scrub growth reappeared in some of the older production zones, and the landscape took on a semi-abandoned appearance in some places (Zweifler, Gold, and Thomas, 1994).[4] In other, more densely settled places, part-time farmers began planting trees for timber production on their smallholdings (Rocheleau et al., 1997). Small forests began to reappear in the Dominican landscape.

The origins of these differences in the historical trajectories of forest cover change are economic. In 1999, the region contained the three poorest peoples in the Western Hemisphere, Nicaragua [$410 gross domestic product (GDP) per capita], Haiti ($450 GDP per capita), and Honduras ($760 GDP per capita). It also contained three of the wealthiest populations in Latin America (in Puerto Rico, Trinidad and Tobago, and Mexico), as well as small but wealthy island populations in the Caribbean (Barbados, the Bahamas, and the Cayman Islands). These contemporary contrasts between poverty and wealth testify to local differences in economic histories that largely account for the contrasting histories of forest destruction and regeneration in the region.

The historical differences in economic conditions translate into differences in the array of economic opportunities available to individuals. From a limited set of choices, people fashion livelihoods as farmers or rural/urban migrants that have consequences for a region's rain forests. These paths through the life course depend in part on the paths chosen by older family members, so the choices available to young adults change from generation to generation in a path-dependent way. Four distinct paths have shaped recent forest cover change in the region. The first path has an endogenous origin: smallholders produce cash crops for expanding numbers of consumers in nearby cities. A second path involves agroexporters and land use changes that competition in worldwide markets for tropical commodities has forced on growers. A third path, associated with grinding poverty, produces poor migrant labor and landscapes with an abandoned look. A fourth path leads through tourism to ecotourism, land abandonment, and forest regeneration. Each path is outlined here.

Expanding Urban Markets and Intensifying Pressures on Forests

During the last two decades of the twentieth century, the political-economic supports for tropical deforestation in Central America and the

Caribbean changed in significant ways. The urban population, located primarily in the highlands or on the dry Pacific coast of the isthmus, increased 79 percent between 1980 and 1999. The corresponding growth in urban consumer demand for dairy and meat products spurred landowners on the humid Atlantic side of the isthmus to convert rain forests into pastures. Similar incentives pushed the landless poor in Honduras up the hillsides of the Cordillera Nombre de Dios and east into the forests of Mosquitia in search of unclaimed lands to farm (Edelman, 1995; Humphries, 1998). Once they cleared these lands and formed small communities, the settlers became part of pressure groups that have lobbied for the extension of roads to the new communities. Once constructed, the new roads generated another wave of land clearing nearby. The growing numbers of urban consumers also created more demands for fuelwood close to the cities. Collectors degraded woodlands in the search for fuelwood outside cities such as Managua or Kingston (Utting, 1993; Tole, 2002). In this manner, the growth of internal consumer demand, tied in part to population growth in the region, generated new economic pressures for continued rain forest destruction and degradation.

Agricultural expansion at the expense of rain forests characterizes places such as Honduras with abundant amounts of forested land, as well as places such as Haiti with very little forested land. Several factors make the exploitation of forested land preferable to the intensification of already cultivated land through technological innovation, even when the forested land may be far from urban markets. Because smallholders in the poorer countries often do not have livestock to provide manure or capital to pay for chemical fertilizers, they do not have the means to replenish nutrient-depleted fields. The absence of secured property rights in their lands also discourages farmers from investing in their cultivated lands. Under these circumstances, they take advantage of the "subsidy from nature" provided by the accumulated fertility of rain forest soils by converting forested lands into fields (Hecht, Anderson, and May, 1988). The continued salience of the POV (poverty) term in the regional QCA for the 1990s (see table 3.1) reflects the persistence of the poverty-induced environmental degradation in the region's poorest nations (Haiti, Honduras, and Nicaragua).

Agro-Exporters, Contract Farming, and Environmental Certification

The easy access to tropical rain forests from the sea made the Caribbean one of the first places in the world where settlers converted tropical rain forests into fields to produce crops for sale overseas. Later, when cultivators in other tropical regions began to grow bananas, coffee, and sugar cane, they frequently became the low-cost producers. To compete with producers in the new regions, longtime producers in the old export enclaves around the Caribbean had to establish the superior quality of their products (Grossman, 1998). For example, their bananas might arrive in Europe with fewer blemishes than the less expensive bananas from Ecuador because, at least prior to the 1960s, West Indian fruit did not have to travel as far to a port of embarkation as Ecuadorian fruit (Wunder, 2001).

The need to enhance the quality of agro-exports linked producers to consumers through the offices of agro-exporters, and it gave the exporters a rationale for imposing changes in the production process. For example, exporters began to insist on particular, labor-intensive procedures in the packing of bananas and more precise applications of fertilizers, all in an effort to increase profits through enhanced productivity or reduced waste (Grossman, 1998:124–208). To enforce their wishes, the agro-exporters began to promote contract farming. The contracts focused the energies of producers on changes in the production process in old fields. In so doing, they de-emphasized agricultural expansion as a means to increased profits. Accordingly, the acreage under banana cultivation in the region has remained relatively static, declining in the Windward Islands (Grossman, 1998:151) and increasing in the Caribbean lowlands of Costa Rica (Vandermeer, 1996:227).

Agro-export expansion explains the early waves of deforestation in the region, but its salience as a driving force has declined since 1980. The validity of this conclusion varies from commodity to commodity. In some places, the old pattern persists. The increasing taste for gourmet coffees in the United States and Europe during the 1980s and 1990s induced smallholders in Jamaica's Blue Mountains to expand the extent of their coffee groves at the expense of the surrounding rain forest (Weis, 2000). Figure 3.2 provides a snapshot of the Blue Mountains in the early 1980s.

The search for competitive advantages by producers in long-settled tropical places has pushed a small number of agro-exporters in the direction of environmental certification schemes during the past 10 years. Certification represents a special case of contract farming. The grower cultivates a crop under specialized conditions in return for a green label and a competitive advantage in overseas markets. By reducing the proportion

Figure 3.2   The Blue Mountains of Jamaica, the recent site of extensive deforestation to cultivate gourmet coffee for export. Photo credit: F. Mattioli, FAO photo (1984).

of the forest cut each year, the owners of a teak plantation in Costa Rica received certification for practicing sustainable forestry, which in turn entitled them to sell their timber for much higher prices in the Dutch market.[5] Coffee is cultivated without pesticides in southern Mexico and sold as certified organic coffee in North American markets. These certification schemes bring environmental benefits to producing areas. The shaded organic coffee fields protect the plant more effectively against pests than the sunny fields of a coffee monoculture, and they also support a greater diversity of arthropods (beetles, ants, and wasps) than the sunny fields (Vandermeer and Perfecto, 1995:136–140). The proportion of Central American forests managed by certified sustainable forestry techniques remains very small (Eba'a Atyi and Simula, 2002), and the extent of organic coffee cultivation remains unclear, so the magnitude of the environmental benefits from certification schemes remains open to question.

Taken together, these trends indicate that fluctuations in consumer demand for agricultural products in the wealthier nations continued to influence land-cover changes in Central America and the Caribbean during the 1980s and 1990s, but these trends no longer drive tropical rain forest destruction in the unambiguous way that they did prior to 1980. A historical turning point has occurred, and more varied trajectories of land use change have begun to characterize the region.

Rural Poverty, Labor Migration, and Forest Recovery

Throughout the twentieth century, poor rural peoples living in the highlands of Latin America have made a living through temporary labor migration. Peasants from the mountainous interior of Puerto Rico would migrate down to the coast to work in sugar cane plantations in the early 1900s (Bergad, 1978). After World War II, both Mayan and mestizo smallholders in the Central American highlands migrated to the Pacific coast to take jobs in the new cotton plantations. More recently, poor urban residents have left the cities for seasonal employment on plantations (Williams, 1986:60–65). By the late twentieth century, looking for work in faraway places had become an accepted cultural practice, a strategy that young adults expected to use to earn a livelihood.

The recent improvements in transportation that have enabled businessmen to create international commodity chains for fruit have also lengthened the distances that migrants will travel to work. People whose parents traveled 50 miles to the Pacific coast of El Salvador to find work 30 years ago now travel 1,500 miles to the southern United States to find work. The numbers of people engaged in temporary labor migration is particularly high in regions such as Central America and the Caribbean, where large concentrations of poor people live a long bus or boat ride from high-wage labor markets in the United States. Labor migrants stream through the region's airports, bus stations, and ports. Mexicans, Haitians, and Dominicans travel to the United States; Jamaicans go to England; Haitians migrate to the Dominican Republic; and Nicaraguans travel to Costa Rica.

In some instances, the departure of large numbers of workers changes the landscape. The uplands of Puerto Rico provide an early historical example of this sequence of cause and effect in landscape change. In the mid-twentieth century, the hillsides of interior Puerto Rico contained extensive groves of shade-grown coffee trees whose harvest required large amounts of labor over several months each year. After 1950, emigration from these hillside communities to the United States and to Puerto Rico's coastal cities gradually deprived landowners of their labor force, and

Table 3.2
*Afforestation Processes in Central America and the Caribbean:
A Qualitative Comparative Analysis (9/9)*

affpol smallag OUTMIGR NONFARM (1) +
affpol smallag PRICEDEC NONFARM (2) +
AFFPOL PRICEDEC OUTMIGR SMALLAG nonfarm (1) +
affpol PRICEDEC OUTMIGR smallag (1) +
AFFPOL pricedec OUTMIGR smallag nonfarm (1) +
AFFPOL pricedec outmigr SMALLAG nonfarm (1)

The "+" in this table indicates that afforestation occurs whenever any of the listed combinations of causal conditions exists.

Upper case indicates the presence of a factor; lower case indicates its absence. The ratio in parentheses indicates the proportion of studies for that period whose findings agree with the Boolean expressions in the lines below it. The number in parentheses after each expression indicates the number of studies whose findings were consistent with the configuration of causes in that expression.

Factors: AFFPOL = government programs to encourage afforestation; NONFARM = large numbers of people have found work in the nonfarm sectors of the economy; OUTMIGR = large numbers of working-age people have left the community in search of work; PRICEDEC = the prices for the most prevalent agricultural crop in the region have experienced a sustained decline; SMALLAG = small-scale agriculture predominates in the place.

they responded by abandoning their land. Now, secondary forests have supplanted coffee groves throughout the humid uplands of the island (Rudel, Perez-Lugo, and Zichal, 2000).

A similar dynamic may have begun in other places in the region. Table 3.2 reports the results of a qualitative comparative analysis of case studies that have recently reported net increases in forest cover.[6] These regions have experienced either extensive out-migration by working-age adults or declines in the prices of the chief agricultural commodity produced in the region. By making agricultural labor more expensive or the income from harvests smaller, both trends diminish the viability of agricultural enterprises and encourage land abandonment. The spontaneous afforestation reported in the coffee-growing regions of western Honduras between 1987 and 1996 may stem from labor scarcities brought on by depressed coffee prices and out-migration from the region (Southworth and Tucker, 2001). Trends in southern and central Mexico, where smallholders engage in corn cultivation, recall the Puerto Rican experience. Labor migration to the United States, newly created jobs in oilfields, and more lucrative handicraft work have taken people away from their corn fields, and secondary forests have increased in the region (Klooster, 2000; Collier, Mountjoy, and Nigh, 1994).

In regions where people earn incomes from cattle rather than from coffee, the labor migration–landscape change dynamic has unfolded in a different way. After 1980, new American import restrictions on beef, coupled with declines in consumer demand for beef in the United States, made cattle ranching less profitable in a number of Central American countries (Edelman, 1995:27–31). In response to declines in cattle prices, many ranchers in long-settled regions of northern Costa Rica allowed some of their lands to revert to forest and devoted more of their labor and capital to other enterprises (Berti Lungo, 1999). Under these circumstances, brush began to encroach on cattle pastures in some long-settled areas. The landscape took on a "haunting semi-abandoned quality, ... with mixed brush and pastures stretching as far as the eye can see" (Edelman, 1995:40–41).

In more remote, newly settled places, in southern Nicaragua, for example, landowners had fewer alternative economic opportunities, so they continued to convert rain forests into pasture even in the face of declines in the price of beef (Larsen, 2000; Kaimowitz, 1996). Because cattle ranching requires extensive amounts of land and little labor, the departure of workers for better-paying jobs elsewhere does not threaten the viability of local agricultural enterprises (Preston, 1998). After the departure of the male family member, children and wives tend to the cattle. They may not devote as much time to either the cattle or the pastures, so the landscape may tend to take on the abandoned appearance noted earlier, but the pastures and the cattle, in smaller numbers, persist.

Poverty, Tourism, and Forest Regeneration in the West Indies

The emergence of nonagricultural economies in the West Indies during the second half of the twentieth century set some islands on a path that prevented further rain forest destruction, but other islands continued down a path of environmental degradation. Decisions about political independence, made in the 1950s and 1960s, played an important role in the subsequent direction of landscape change. Some islands remained political dependencies of European and North American powers, whereas others became independent states. Outside investors chose to locate tourist facilities in the politically dependent islands and avoided the independent islands. The political stability and subsidies provided by the colonial powers, coupled with direct air links to metropolitan centers in North America and Europe, made the dependent islands more attractive sites for constructing tourist facilities (Bates, 2001).

Because tourist enterprises require large amounts of labor, they competed with plantations for workers, and the increased labor costs on plantations accelerated agricultural decline on islands with emerging tourist economies. Although the beaches and warm climate attracted the tourists, accessible spots of undisturbed tropical rain forest added to the islands' natural attractions, so forest preservation acquired a practical value in tourist oriented island economies. Planners and proprietors of tourist enterprises lobbied for the creation of parks to preserve forested glens and ridges on the islands. Their success led to extensive park systems on some islands—for example, parks occupy over 70 percent of the island of Martinique; with these changes, forest cover has stabilized on the wealthier islands. The contingent nature of these environmental gains, the way that they depend on fluctuating local circumstances, is well illustrated by the story of Hortense Welch (see box). On the poorer, politically independent islands, agriculture remained the predominant economic activity, and smallholders usually sought to improve their economic prospects by converting forested hillside slopes into banana plantations (Harcourt and Sayer, 1996:131,136). Table 3.3 summarizes these patterns.

Conclusion

The four paths outlined here predominate to varying degrees in different places. The third path of poverty and out-migration typified Puerto Rico from 1950 to 1980; it now describes rural places in El Salvador (Hecht, 2004). A mixture of the first path, urban expansion, and the third path, rural poverty and out-migration, describes land-use trends in contemporary Nicaragua. Combinations of sometimes unrelated historical events and ecological conditions set places on particular paths of landscape change. For example, the out-migration and reforestation of Puerto Rico after 1950 occurred only because at the end of the nineteenth century the United States occupied an island containing large numbers of poor people who earned their livelihoods by working marginal lands in a labor-intensive way. In the Lesser Antilles, forests expanded during the 1990s on islands with burgeoning tourist trades, which in turn had their origins in decisions by some colonial era elites to remain political dependencies rather than become independent nations after 1960. In both Puerto Rico and the Lesser Antilles, seemingly unrelated political events shaped subsequent land use changes, creating path dependencies.

Table 3.3

Poverty, Tourism, and Forest-Cover Change in the Lesser Antilles *

	Politically Independent (%)	Tourism Receipts per Capita, 1997	Gross Domestic Product (GDP) per Capita, 1995	Annual Forest-Cover Change, 1990–2000 (%)
Wealthy Islands†	33	$5,769	$12,770	+0.233
Poor Islands	83	$1,018	$3,670	−1.1

SOURCES FOR DATA: GDP, United Nations, 1997 Statistical Yearbook, New York; tourism receipts per capita, United Nations, 1997 Statistical Yearbook, New York, and United Nations, 1997 Demographic Yearbook, New York; annual forest-cover change, Forest Resources Assessment 2000, available at www. fao.org/forestry/fo/fra/index.jsp.
 * All mean differences are statistically significant at levels of *P* < .1.
 †Wealthy islands include Antigua, Bahamas, Barbados, British Virgin Islands, Cayman Islands, Guadeloupe, Martinique, Netherlands Antilles, and U.S. Virgin Islands. Poor islands include Dominica, Grenada, Montserrat, Saint Kitts and Nevis, Saint Lucia, and Saint Vincent and the Grenadines.

Although demand from urban consumers within the Central American and Caribbean region drives more deforestation now than it did in the mid-twentieth century, external markets, politics, and social networks, established over the past four centuries, continue to exercise an extraordinary influence over land use. Enclave agro-export economies with ties to consumers in affluent countries, out-migration to labor markets in nearby affluent countries, and links in the tourist trade to former colonial powers shape the status of tropical rain forests in the region. All of these institutional arrangements originated in earlier periods of economic and political domination by northern nations. Northern environmental movements perpetuate this pattern of influence, and, more so than in other tropical rain forest regions, they have enhanced rain forest preservation by promoting ecotourism and environmental certification. Although the influence of nongovernmental organizations over rain forest conservation in Central American and the Caribbean is visible, it is not unlimited. These organizations cannot affect the growth of local urban markets for tropical commodities, or the departure of young workers for distant labor markets.

The economic growth of the ecotourist sector in the regional economy makes the prospects for integrated conservation and development projects (ICDPs) better in Central America and the Caribbean than elsewhere in the tropics. Conservationists usually try to initiate ICDPs in the buffer

---

**Hortense Welch: The Changing Fortunes
of an Ecotourism Entrepreneur**

Born in the early 1960s, Welch lived at home in Gales Point Manatee, Belize, with her partner (Moses Andrewin) and five children in the early 1990s. Gales Point is located on a lagoon adjacent to a manatee reserve, 4 hours by boat south of Belize City, the major port of embarkation for travelers to Belize. During the early 1990s, Gales Point became a destination for ecotourists visiting Belize. Moses had been providing for the household by hunting and selling bushmeat and by making periodic trips abroad to work. Hortense supplemented their income through the sale of cakes and roasted cashews to local residents. When the number of ecotourists passing through Gales Point on their way to the manatee reserve began to increase in the early 1990s, Moses gave up hunting and began to serve as a guide for birdwatchers, tour boat operators, and visitors to the Mayan caves. Hortense set up a bed and breakfast (B&B) business to serve the tourists. With her newfound interest, she became head of the Gales Point B&B Association and an influential participant in the community ecotourism project. Other local residents soon became envious of Hortense's business successes. After several years of simmering conflict and the end of government support for the development of the tourist trade, the community ecotourist association collapsed. Without the ecotourist organization to recruit tourists and transport them to Gales Point, the number of tourists declined during the late 1990s. At the same time, illegal loggers began to work inside the Manatee reserve. By 1999, Hortense reported herself to be "burnt out" by her ecotourist ventures.

SOURCE: Belsky, 1999; and Belsky, 2002, personal communication.

---

zones around parks as part of strategies to maintain the ecological integrity of the parks. ICDPs have had a troubled history since the late 1980s when they became the most frequently employed approach to tropical conservation (McShane, 1999), but there are several reasons to expect that their success rate would be higher in Central America and the Caribbean than elsewhere. The relatively large volume of tourists who visit Central American and Caribbean parks makes adjacent ICDPs somewhat more viable than they are elsewhere. Jobs in the tourist industry provide livelihoods for many local heads of households and reduce the incentive to exploit forests in and around the parks. Under these circumstances, the ICDPs do not have to provide as many jobs for locals, and they are more likely to induce some recovery of the forests around the parks.

The early and extensive deforestation of the region makes secondary forests that have regenerated on abandoned agricultural lands more common than primary forests. The growth in the extent of secondary forests probably explains why net deforestation has begun to decline in the region. By the mid-1990s, Costa Rica had more secondary than primary forests, and deforestation continued, albeit at a slower rate. It declined from 2.8 percent per annum in the 1980s to 0.8 percent per annum in the 1990s (FAO, 1993, 2001). In this context, policies that promote the regeneration of rain forests, such as the Clean Development Mechanism of the Kyoto Protocol, could accelerate the expansion of secondary forests in the region.

The Amazon Basin: The Breakdown of Passive Protection

The Primeval Forests
… There were enormous trees, crowned with magnificent foliage,
decked with fantastic parasites, and hung over with lianas which varied
in thickness from slender threads to huge python-like masses, … now
round, now flattened, now knotted, and now twisted with the regular-
ity of a cable. Intermixed with the trees, and often equal to them in
altitude, grew noble palms; … other and far lovelier species of the same
family, their ringed stems sometimes scarcely exceeding a finger's thick-
ness, but bearing plume-like fronds and pendulous bunches of black
and red berries, quite like those of their loftier allies, formed along with
shrubs and arbuscles of many types, a bushy undergrowth, not very dif-
ficult to penetrate.
RICHARD SPRUCE, *Notes of a Botanist on the Andes and the Amazon* (1908:17)

An immense tropical rain forest covers the Amazon and Orinoco River basins (figure 4.1). Extending from the Andes to the Atlantic Ocean, the Amazon–Orinoco forest contains a little less than half of all of the forested land in the tropics.[1] Despite the forest's large size, humans have made sustained efforts to exploit the lands beneath it. Amerindians estab-

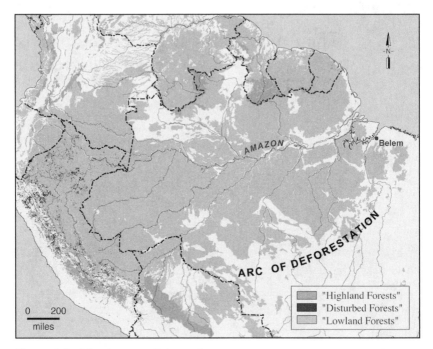

Figure 4.1   The Amazon–Orinoco forest.

lished large settlements and cut extensive fields out of the floodplain forests along the main stem of the Amazon river during the pre-Columbian era (Denevan, 1992). Amerindians also established settlements and cultivated manioc away from the river in areas such as the Llanos de Mojos near the headwaters of the Madeira River, a tributary of the Amazon (Cleary, 2001:75–79). Diseases and slave traders began killing Amerindians almost immediately after contact. By the seventeenth century, secondary forests had begun to reoccupy the depopulated riverine landscapes. The nineteenth- and early twentieth-century rubber boom did not alter the landscape, so the Amazon–Orinoco forest had probably expanded to its greatest extent in centuries when the current wave of deforestation began after 1960 (Cleary, 2001).

During the past 15 years, concern about the destruction of the Amazon forests has prompted a series of efforts by both international and Brazilian scientists to measure the loss of forest through remote sensing techniques. Since 1988, the Brazilian space agency has used satellite images to measure yearly fluctuations in the rate of deforestation. Be-

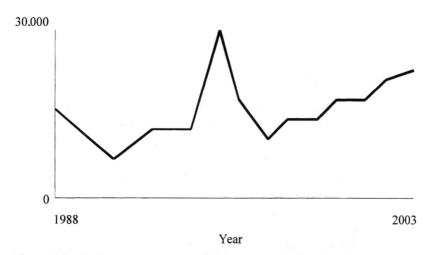

30.000

0

1988   2003

Year

Figure 4.2   Deforestation (in square kilometers) in the Brazilian Amazon, 1988–2003. Source: Instituto Nacional de Pesquisas Espacials (2004) (adapted from information available at http://www.obt.inpe.br).

cause Brazil contains approximately two thirds of the Amazon forest, the year-to-year changes for Brazil, presented in figure 4.2, provide an approximate measure of current, basin-wide trends in deforestation. Since 1998, the annual rate of deforestation for Brazil has averaged around 0.5 percent per year. The Food and Agricultural Organization of the United Nations (FAO) figures for the entire region, presented in table 1.1, indicate a decline in deforestation from 0.7 percent to 0.4 percent per year from the 1980s to the 1990s. The decline probably resulted in part from initial overestimates of the deforestation rate for the 1980s and in part from real declines in the deforestation rate in the upper reaches of the Amazon basin, in particular in Ecuador and Colombia. The relatively low rates of deforestation in the basin should not give us cause for complacency. Each year, people destroy more rain forest in the Amazon Basin than in any other region of the tropics. They also degrade large tracts of forest through logging and fires. Projected over decades, this path of destruction would cause globally significant losses of biodiversity and accelerate climate change through the release of large amounts of carbon dioxide into the atmosphere. Its origins lie in a historical sequence of state-sponsored development followed by enterprise-driven land-use conversion during the last three decades of the twentieth century. In the following pages, I outline the setting, the subregional variations, and the changing causes for forest cover change in the Amazon basin.

The Forest and the People

The Amazon–Orinoco basin contains the world's largest block of tropical rain forest; these forests extend approximately 3,400 km east to west, from the mouth of the Amazon river to the eastern slope of the Andes, and 2,400 km south to north, from the savannas of the Brazilian *sertao* to the shores of the Atlantic Ocean in the Guianas and eastern Venezuela. The large size of the forest conjures up images of a uniform, forested plain extending as far as the eye can see, but in reality the forest covers a wide variety of landforms. Andean foothills in the east, stone-towered table mountains (tepuis) in the north, and mineral-rich ridges in the southeast all rise out of the forested Amazon landscape. As a traveler moves away from large rivers, the *varzea*, the fertile alluvial plains, gives way to *terra firme*, the extensive upland regions between the rivers. Soil resources in the *terra firme* vary over short distances, sometimes from the top to the bottom of a hill, but infertile soils, oxisols, predominate in most regions. In some regions, such as Rondônia and Para, oxisols alternate with more fertile alfisols, otherwise known as *terra roxa* soils (Eden, 1990:30–37).

The climate also varies across the basin. As moist ocean air moves from east to west across the region, the rising land masses in the west extract the moisture from the skies, making western portions wetter than eastern portions of the watershed. Rainfall in the east also shows a marked seasonality that is absent in the west. The variations in rainfall have shaped plant communities in the basin. Savannas mix with forests in the southern and eastern portions of the basin, which have pronounced dry seasons. In the more humid, western and northern portions of the basin, the abundance of niches and rainfall where the Andes rise out of the Amazon plain have contributed to the world's most diverse assemblage of vascular plants. The tropical edges of the Andes contain approximately 45,000 known plant species, of which 20,000 are endemic to the region (Mittermeier et al., 2000:37–38). Even disturbed habitats contain diverse plant communities. Transects, totaling 1 hectare in extent, running through secondary forests at the eastern base of the Andes in Ecuador, contained 121 tree species (Rudel, Bates, and Machinguiashi, 2002a)!

The people who live in and around the Amazon basin are poor, but they are not among the world's poorest people. The World Bank categorizes the Amazon basin countries as middle income countries, ranging from lower middle income countries such as Ecuador (2001 gross national product [GNP], $1,360) and Bolivia (2001 GNP, $990) to upper middle income countries such as Brazil (2001 GNP, $4,350). Although the people of Amazon basin countries do not begin to approach the industrialized European, North American, and East Asian peoples in wealth, they are much wealthier than southern Asian and African peoples. For example, in 2001 in parts of the Ecuadorian Amazon, wage rates for unskilled labor averaged $10 per day, compared with $1 to $2 a day in Zimbabwe.[2] Most of the people who inhabit the Amazon basin live in cities. Urban residents constituted 78.3 percent of the population in the Amazon basin countries in 1999 (World Bank, 2001).

The aggregated data fail, however, to convey the severity of the poverty experienced by many rural people in the Amazon basin. The distributions of land and income are among the most unequal in the world. For example, in the mid-1990s, Brazil had more than 1.25 million landless peasant households in the same agricultural sector that contained 21,000 properties with more than 2,000 hectares of land (IBGE, 2001). Although the wealthiest 20 percent of Brazilian households earned 63 percent of the national income, the poorest 20 percent of the households earned only 2.6 percent of the national income in 2000 (World Bank, 2001).[3] These inequalities have consequences for forest cover. Companies and wealthy households provide the capital for development projects that break down the region's isolation and create economic opportunities for people who want to convert forests into fields and pastures. Poor people, now mostly from within the region (Vosti, Witcover, and Carpentier, 2002), provide the hard physical labor that converts forests into income-producing lands. Governments assist their citizens in transforming the landscape, with the Brazilian state exhibiting more capacity to do so than the Andean states. Although all of the states with lands in the Amazon launched programs to settle the region during periods of extraordinary state activity in the 1960s and 1970s, only the Brazilian state has persisted with programs of regional development during the more fiscally strapped, neoliberal political conditions of the 1980s and 1990s.

These variations in physical resources, economic potential, and political order make it difficult, but not impossible, to generalize about the paths of forest destruction in the Amazon basin. A brief historical narrative of the events that have driven deforestation in the basin during last 40 years should illuminate how these different elements have combined to create an overall trajectory of change as well as identifiable subregional trajectories of change within the basin.

Passive Protection and Paths of Destruction, 1960–2000

Passive Protections for Nature

For almost all of the nineteenth and twentieth centuries, the inaccessibility of the region provided a kind of passive protection for flora and fauna of the Amazon rain forests. Outsiders could earn profits only from the exploitation of high-value, low-volume products with low transport costs, such as gold from placer deposits in streams and rubber from naturally occurring stands of trees in the forest (Barham and Coomes, 1996; Jaramillo Alvarado, 1936). Bulky commodities such as timber or agricultural commodities cost too much to ship to markets from remote places, so prospective entrepreneurs looked elsewhere to establish farms or sawmills. The region's forests remained largely intact.

Efforts to make the region more accessible encountered formidable natural obstacles. To the west, the *Cordillera Oriental* of the Andes walled off people in the populous highlands from the sparsely populated Amazon lowlands only 100 miles to the east. Elites in the highland centers of the Andean countries frequently expressed a desire to integrate the Amazon lowlands into national society, but these programs invariably required the construction of railroads and penetration roads across jagged mountain ranges and down narrow river valleys to the Amazon plain (Hegen, 1966). The rugged topography, combined with the limited amounts of capital available for road and railway building, ensured that the construction of transportation links to the Amazon progressed slowly, if at all.

Brazilian elites did not have to contend with mountainous barriers, but the long distances between the Amazon basin and Brazil's southern cities, coupled with the coastal orientation of most Brazilian elites, protected the region's forests from extensive exploitation. Until the twentieth century, most Brazilians lived directly or indirectly off the proceeds from export agriculture, usually either sugar cane or coffee. The economic reliance on income from overseas oriented Brazilians toward economic activities along the southern coast of the country. Between the seventeenth and twentieth centuries, settlers decimated the Atlantic forest to establish plantations, while the Amazon basin, 1,500 miles to the north, remained inaccessible, out of sight and out of mind (Dean, 1995). Without paved roads to shorten the transportation times to markets in southern Brazil, truckers could not move Amazon products to domestic markets. The seas did not provide an alternative route to markets given the lengthy ocean route from Belem at the mouth of the Amazon around Cape Sao Roque to urban centers in the south.

Since 1965, a series of large infrastructure projects have reduced the distance between the Amazon and urban markets in the south, but passive protections continue to preserve forests in the northwestern portions of the basin. Between 1991 and 1996, 82 percent of all deforestation in the Brazilian Amazon occurred in three provinces—Rondônia, Mato Grosso, and Para—within the "arc of deforestation," depicted in figure 4.1, along the southern and eastern edges of Amazonia (Alves, 2002). Areas to the north and west of the arc are too far from urban markets in southern Brazil to sustain commercial agriculture, so deforestation there occurs more slowly. Deforestation occurred at 1.5 percent per annum between 1989 and 1997 in the three "arc" provinces, and, immediately to the north and west, in Acre, it occurred at only 0.5 percent per annum during the same period (Vosti et al., 2001:116).

Marketing constraints have continued to handicap agricultural producers along the western edges of the basin. Despite suitable agro-ecological conditions and a wealth of experience working on coastal banana plantations, smallholders in the southern Ecuadorian Amazon cannot cultivate bananas commercially because the cost of transporting bananas over the Andes to ports on the Pacific is too high. Not surprisingly, the deforestation rate in Ecuador is lower east of the Andes than it is to the west. It was 0.6 percent per annum east of the Andes between 1983 and 1993 compared with 1.95 percent per annum west of the Andes between 1986 and 1996 (Sierra and Stallings, 1998; Sierra, 2000).

Climatic conditions also protect the northwestern rain forests. These regions receive more than 2,200 mm of rainfall per year and have no pronounced dry season. Under these conditions, farmers find it very difficult to practice modern agriculture. The heavy rains wash nutrients from the soils, so soil fertility is usually low. The absence of a dry season allows pests to multiply and infest crops. The humid conditions also make it difficult to use agricultural machinery because the equipment becomes mired in the mud. Logging and ranching remain possible, but the low fertility of the soils limits the intensity of livestock operations. Taken together, these conditions make it difficult to continue farming cleared land, so people

frequently allow fields to revert to forest in the older zones of colonization (Schneider et al., 2000; Rudel, Bates, and Machinguiashi, 2002a;
Maki, Kalliola, and Vuorinen, 2001). Together with the long distances to
market, these climactic conditions protect nature passively; humans do
not do anything to preserve the forests, but they continue to exist.

Large Projects Destroy Passive Protections, 1960–1980

Beginning in the 1960s, a series of expensive projects initiated by governments and corporations began to break down the isolation of the Amazon
and facilitate the destruction of its forests. In many instances, government
elites initiated these projects for geopolitical reasons. They wanted to tie
the remote, sparsely populated Amazon region more securely to centers
of power in the societies that they governed. The Brazilian generals christened their first effort the Programa de Integracao Nacional (PIN). It
featured extensive road building, and then land settlement schemes along
the newly constructed highways. Other regional development efforts followed in quick succession. The Brazilian military replaced PIN with Po-
loAmazonia, which poured state funds into "growth poles" of concentrated economic activity, such as the huge iron ore mines at Carajas in the
eastern Amazon. In the mid-1980s, the Polonoroeste project used World
Bank funds to pave roads and subsidize prospective farmers in Rondônia,
along the southwestern margins of the forest. In the late 1980s, the Calha
Norte project increased the military presence and accelerated highway
and dam construction efforts in the Guiana region north of the Amazon
river (Hecht and Cockburn, 1989:95–128). In the late 1990s, Brazil's
government launched another ambitious initiative, Avanca Brazil, which
promised to bring producers in the interior portions of the basin closer
to markets inside and outside of Brazil through an ambitious program of
road paving and construction (Laurence et al., 2001).

   Government officials in the Andean countries offered geopolitical reasons for converting Amazon forests into farms. In an effort to prevent
incursions by foreigners, Ecuadorian officials talked about establishing
"living frontiers," a string of new communities in the rain forest, populated by Ecuadorians, along the Amazon borders with Peru and Colombia (Rudel with Horowitz, 1993:60–74). In Bolivia, Peru, and Colombia,
governments launched new land settlement schemes at the base of the
Andes during the 1960s in an effort to secure remote regions and placate
poor, restive rural peoples with free land (Weil and Weil, 1983; Ortiz,

1984; Santos-Granero and Barclay, 1998:66–97). The funds to build the new penetration roads often came from the private sector. Oil companies in Ecuador expanded their road networks as they opened up new areas for exploitation. In most instances, *mestizos* and acculturated Amerindians quickly took advantage of the improved access and established farms along the new roads, creating additional corridors of cleared land in the forest (Sierra, 2000:6).

When rumors circulated about possible new public investments in infrastructure, potential beneficiaries (i.e., large landowners, local businessmen, and smallholders) organized and began to lobby for the construction of new roads. These "growth coalitions" exerted political pressure for both large and small investments in infrastructure. Soybean producers lobbied for the paving of the Cuiaba-Santarem highway, a thousand-mile of stretch of all-weather road that would give them direct access to a deep-water port and markets overseas (Nepstad et al., 2002). Real estate developers and small farmers in Rondônia lobbied for revisions in the state zoning plan and the extension of feeder roads that would open up more forested areas in the province for conversion to agricultural uses (Mahar and Ducrot, 1998). The political support for Avanca Brazil in the Brazilian national congress came, and still comes, from an alliance of local growth coalitions. Both individually and collectively, these coalitions have shaped the decision making in Brasilia about Amazon transportation and land use.

Other political and institutional factors accelerate land clearing in the frontier regions of the Amazon. In particular, the frequent inability or unwillingness of the central governments in the Amazon basin to establish and defend titles to land in frontier regions creates an informal system in which "he who works the land, owns it." The construction of each new stretch of a penetration road makes new tracts of untitled and forested land accessible to urban markets and triggers a rush to establish claims to land by working it. The rush to claim land by clearing it eventually subsides, but only after all of the land has claimants who physically occupy it (Rudel, 1995). If, however, the state never officially recognizes these claims, the local social order that protects claimed but uncleared tracts of land can disintegrate. Political turmoil in the provincial and the federal governments also upsets the informal order and encourages a new round of land conflicts in which people assert claims to land by clearing it.

Many of the smallholders who work to establish claims to land at the forests' margins do not have enough capital to establish commercially

Table 4.1
Deforestation Processes in the Amazon Basin:
A Qualitative Comparative Analysis

A. 1970s and 1980s (30/32)
SMALLAG COLON (13) +
SMALLAG ROADS (7) +
RANCHING SMALLAG (2) +
ranching ROADS (2) +
ROADS colon (4)

B. 1990s (32/32)
SMALLAG ROADS (12) +
RANCHING ROADS (3) +
ROADS logging (7) +
RANCHING SMALLAG LOGGING (5) +
RANCHING smallag logging (2)

The "+" in this table indicates that deforestation occurs whenever any of the listed combinations of causal conditions exists.

Upper case indicates the presence of a factor, and lower case indicates its absence. The ratio in parentheses indicates the proportion of studies for that period whose findings agree with the Boolean expressions in the lines below it. The number in parentheses after each expression indicates the number of studies whose findings were consistent with the configuration of causes in that expression.

Factors: COLON = agricultural colonization of new regions with government assistance; LOGGING = timber exploitation; RANCHING = cattle ranching; ROADS = construction of penetration roads into roadless regions; SMALLAG = smallholder agriculture.

successful farms, so, having cleared an appreciable portion of their land, they sell out to larger, better-financed landowners. This system, often referred to as the *colono* system, characterized frontier agriculture in Latin America throughout the twentieth century. It continues to characterize places such as Sao Felix in Para, where poor people, sometimes referred to as spontaneous colonists, start farms at the forest's edge with no government assistance (Maturana, 1999; Mertens et al., 2002).

The Retreat of the State from Rural Areas, 1980–2000: A Qualitative Comparative Analysis

Although large projects, growth coalitions, and weak land tenure institutions shaped deforestation processes throughout the late twentieth century in the Amazon basin, some significant changes in these processes did take place in the 1980s and the 1990s. The qualitative comparative analysis presented in table 4.1 outlines these changes. Parts A and B of table 4.1 present the configurations of causal forces in the literature that drove deforestation in the 1970s and 1980s (part A) and in the 1990s (part B). Perhaps the most significant difference between the early and later descriptions of the causal forces supporting deforestation is the relative absence of the public sector in the later period.

The "colonization" term, present during the earlier period, drops out of the 1990s descriptions of deforestation processes; this change signals the retreat of the public sector from a leading role in deforestation processes during the 1990s. By the late 1980s, the Andean governments, operating under debt-induced plans for fiscal austerity, had scaled back or eliminated colonization programs in the upper reaches of the Amazon basin. These governments no longer built penetration roads to open up new areas of the Amazon basin for exploitation, but they continued to build short, feeder and farm-to-market roads in response to political pressures by local landowners. Government settlement schemes continued in Brazil, but on a smaller scale than previously. INCRA, the Brazilian colonization agency, continued to support the expansion of settlements in Rondônia and Para, but it did not, as it did in the 1970s and 1980s, initiate new settlement schemes in parts of the Amazon where it had previously not been active (Maturana, 1999; Mahar and Ducrot, 1998). Similarly, the Brazilian state in the 1990s did not build new penetration roads, such as the Transamazon highway, for entirely geopolitical reasons. Increasingly, deforestation processes appeared to be enterprise driven. Expansion plans or changes in operations by mineral companies, logging firms, ranchers, and soybean producers drove deforestation during the 1990s. The appearance of the "logging" term in the QCA for the 1990s (part B of table 4.1) signifies this shift in the forces driving deforestation. The Brazilian state continues to play a leading role in the process, but only in concert with coalitions of entrepreneurs who see commercial advantages in major new infrastructure projects such as the paving of the Cuiaba-Santarem highway or the creation of an Araguaia-Tocantins waterway.

Although the salience of colonization schemes has declined over the past 20 years, the importance of road paving projects has increased. When construction crews pave roads, they convert the roads from thoroughfares that frequently become impassable during wet seasons into all-weather roads that provide constant access to distant markets. For producers with perishable products, such as milk, road paving marks a crucial change in their access to markets. Dairy farming really became possible in Rondônia only in the mid-1980s, when the paving of BR-364 made it possible to ship dairy products from Rondônia to the cities of southern Brazil. Road paving also facilitates high-volume operations that produce a constant stream of products; for this reason, large operations involving logging, ranching, and soybean production have lobbied for the paving of roads in Brazil (Nepstad et al., 2002). Not surprisingly, road paving also accelerates deforestation in the vicinity of the recently improved roads. An analysis of road paving in the Brazilian arc of deforestation indicates that, whereas farmers cleared 0 percent to 9 percent of their land within 50 km of unpaved roads, they cleared from 29 percent to 58 percent of the land within 50 km of paved roads (Nepstad et al., 2001).

The reduced government participation in the development of the Amazon basin had some unintended effects. When Brazilian politicians abolished subsidies for large-scale land clearing in the late 1980s, they inadvertently stimulated logging in the Brazilian Amazon. Deprived of state subsidies, large landowners began to log tracts of forest before they converted the forests into pasture. The revenues from the sale of logs then financed the clearing of land and the initial purchase of cattle.[4] To process logs, other entrepreneurs established sawmills along active frontiers, and the presence of sawmills stimulated more logging, at least until loggers exhausted local supplies of commercially valuable timber.

The growth in logging increased the susceptibility of forests to fire because loggers leave behind fuel in the form of slash (deadwood) and because, by reducing canopy cover, they dry out the remaining forest. Once burned, forests become more susceptible to further burning. Through this process, the frequency of fires in the drier, eastern portions of the basin has increased to one fire every 7 to 14 years. Fires at such a high frequency convert forests into scrub or grasslands over time. In two regions south of Belem, 1,280 km^2 and 2,640 km^2 in extent, fires caused more deforestation than land clearing during the mid-1990s (Cochrane et al., 1999; Nepstad et al., 1999). Momentum characterizes this process of landscape change. As enterprise-driven deforestation expands in scale, it gives rise to a new fire regime that destroys additional tracts of forest.

Resource Partitioning and Regional Trajectories of Forest Cover Change Within the Amazon

Do events in the Brazilian arc of deforestation tell a basin-wide story of forest destruction? They do not. People have cleared land in concentrated areas, leaving forests intact, in other parts of the Amazon basin (Alves,

2002). In the Brazilian *varzea*, forest cover may actually be increasing. In the oil- and mineral-rich regions, each new discovery of mineral resources destroys more forest, but the amount of forest lost with each discovery has declined. In old colonization zones, colonists and Amerindians continue to convert small tracts of primary rain forest into fields and pastures.

Differences in the historical legacies of earlier development efforts, coupled with differences in natural resource endowments, account for the contrasting subregional patterns of forest cover change. The Polonoreste project in Rondônia peopled the province with small farmers intent on expanding their income from the new farms. The Carajas project, centered around a multinational conglomerate, the Companhia Valle do Rio Doce (CVRD), transformed a forested landscape into an industrial complex that employed thousands of workers to extract iron ore, process some of it, and ship the rest of it overseas. Struggles between contending groups shaped subsequent patterns of land-use conversion in other places. Faced with colonist invasions, the Shuar, an Amerindian group in the Ecuadorian Amazon, created a political organization and secured about 50 percent of the land in southeastern Ecuador for themselves during the 1970s and 1980s (Rudel with Horowitz, 1993). In northeastern Mato Grosso, land speculators and timber companies expelled a small indigenous group, the Riktbaktsa, from lands along the Cuiaba–Porto Velho road when contractors paved the road during the 1980s (Barraclough and Ghimire, 2000:40–42). Rubber tappers in Acre, led by Chico Mendes, established an extractive reserve in the face of threats by cattle ranchers (Cowell, 1990:169–200). In all of these instances, resource partitioning occurs. People with particular interests secure access to a tract of rain forest, and their interests shape subsequent rates of deforestation on these lands. Where ranchers predominate, deforestation proceeds rapidly; where unacculturated indigenous peoples hold the land, deforestation proceeds more slowly, if at all. In this fashion, a mosaic of land uses has emerged, with each patch in the mosaic exhibiting a different trajectory of land-cover change.

Table 4.2 outlines the largest patches in the mosaic of Amazon land cover. It categorizes the accessible Amazon regions along two dimensions, the value of natural resources in a region and the wealth of the humans who exploit the region's resources. The natural resources in table 4.2 refer to soil as well as mineral resources. Observers have repeatedly noted the differences in soil fertility between the rich, alluvial soils of the *varzea*

Table 4.2

Regional Patterns of Forest-Cover Change in the Amazon Basin

	Rich in Natural Resources	Poor in Natural Resources
Populist Frontier (Capital Poor)	1. *Varzea*—farms with *terra roxa* soils in Brazil Some Amerindian reserves Some reforestation, spreading agroforestry (cacao, coffee)	2. *Terra firme*–parts of Rondônia, Brazil Rapid deforestation–abandonment, out-migration, small dairy and cattle ranches, fire as a management tool, state-subsidized credit
Corporate and Agribusiness Frontier (Capital Rich)	3. Large mineral deposits— Carajas, northeastern Ecuador Slowing deforestation, some ecological modernization Limited access oil fields, with gates or without roads	4. *Terra firme*—Mato Grosso, Brazil Rapid deforestation Soybean cultivation, large cattle ranches, fire as a management tool, state-subsidized credit
Passively Protected Places Beyond the Margins of the Frontier	5. Amazonas, Brazil; northeastern Peru Little deforestation: passive protection supplemented by Amerindian and park ranger protection	

and the comparatively infertile soils of interfluvial regions (Meggers, 1971; Lathrap, 1970). The other dimension in the table refers to differences in the human agents of land-cover change. Corporations have opened up mining regions such as the Carajas for exploitation through expenditures for roads and railroads. Because corporate decisions have proved so important to the subsequent development of these regions, they represent "corporate frontiers." In places such as Rondônia, poor migrant families who carve farms out of the forest constitute the key actors in the development and deforestation of the region. These places represent "populist frontiers" (Browder and Godfrey, 1997). The Amazon basin also contains remote places beyond the margins of any frontier. These regional endowments of people and natural resources represent different starting points in the development and deforestation of regions within the Amazon. The subsequent paths of change in regional land cover—rapid deforestation in one case, afforestation in another case—depend in large part on these initial starting points. I outline these regional paths and path dependencies next.

Figure 4.3 A flooded forest in the Peruvian floodplain of the Amazon River.
Photo credit: A. Brack, FAO photo (1995).

Poor People on Rich Lands: *Varzea, Terra Roxa,* and Forest Cover Change

Poor people work rich lands in some parts of the Amazon basin. *Riberenos,* known as *caboclos* in Brasil, earn their livelihoods from the *varzea,* the band of alluvial lands that border the basin's major rivers (figure 4.3). In some of Brazil's older colonization zones, the first settlers, oftentimes *caboclos,* established farms on fertile *terra roxa* soils. In these resource-rich, densely settled districts, a distinctive pattern of land-use change has emerged during the past 20 years. The rapid growth of cities in the Amazon basin has shaped these changes in land use by expanding the size of markets for *varzea* products. Although the seasonal flooding of the *varzea* limits the production alternatives available to smallholders, they enjoy extraordinary access to urban markets because, with their close proximity to rivers, they can ship their produce to market by water. With growth in the size of markets for wood and agroforestry products in cities such as Belem, Manuas, and Iquitos, smallholders in the *varzea* have recently established small forest plantations (Pinedo-Vasquez et al., 2001).

The growth in urban markets encouraged *caboclos* around Belem to move from collecting acai fruit (hearts of palm) to cultivating it in plantations (Brondizio and Siqueira, 1997). A similar pattern of change may be occurring in the *varzea* forestry sector. Small-scale timber operations, run by *caboclos,* account for most of the timber extracted from the forests of the Brazilian *varzea* (Barros and Uhl, 1995). Recently, smallholders have begun planting and harvesting fast-growing species for sale in urban markets. The planted species include *Calycophyllum spruceanum*, which reaches commercial size in only 8 years (Pinedo-Vasquez et al., 2001). These small-scale agroforestry and plantation forestry endeavors have a high return and do not require full-time labor, which leaves smallholders free to pursue subsistence activities or work in nearby urban centers. The prevalence of agroforestry and the growth in the extent of managed forests in the *varzea* reduced deforestation in these regions to insignificant levels during the 1980s and 1990s (Brondizio et al., 1994; Hiroaka, 1995).

A somewhat similar pattern of land use has developed around Altamira in Para, where smallholders dedicate less fertile lands to pasture and the more fertile *terra roxa* to coffee and cacao cultivation (Moran, Brondizio, and McCracken, 2002). This pattern, coupled with forest cover trends in the *varzea,* suggests that soil fertility may influence forest cover trends in the tropics in ways unanticipated by the forest transition hypothesis developed by Mather for temperate forests (Mather and Needle, 1998). In the forest transition hypothesis, only the most fertile lands remain in agriculture as people give up farming for urban pursuits. The pattern here suggests that the most fertile soils tend to remain forested because they support highly productive agroforestry systems better than do less fertile soils. Cultural factors also sometimes play a role in the emergence of agroforestry landscapes. Some Amerindians show a preference for agroforestry, and in some instances in the Ecuadorian Amazon they have persisted in practicing agroforestry, even on lands with poor soils (Rudel, Bates, and Machinguiashi, 2002b).

Poor People on Impoverished Soils: State Subsidies and Passive Protection of Forests on Populist Frontiers

Along the populist frontiers of Amazonia, smallholders frequently face a dilemma when they carve a small farm out of the forest. They have expended all of their capital on acquiring a farm and settling their family on it, so they have only their labor to use in developing their farms. Typically, smallholders begin by clearing small amounts of land to grow food crops, and, when the yields of these lands decline after two or three harvests, they convert the land into pasture. This process of land conversion occurs in small increments, affecting 1 or 2 hectares per year (Fugisaka et al., 1996). It occurs when the head of the household has accumulated enough of a cash reserve to allow him to forsake wage labor for several months and work on his own lands. Under these circumstances, much of the land on smallholder farms remains forested. Most smallholders in the Ecuadorian Amazon, for example, still had more than half of their land in forest 10 years after they settled in the region (Pichon et al., 2001:157).

Many smallholders also have to contend with isolation and poor soils. As the most recently arrived colonists in an area, they often occupy tracts of land on impoverished soils, located far from roads. Low yields and the expense of getting products to market from these isolated locations cause many smallholders to abandon their farms or sell them to large scale cattle ranchers after several seasons. When smallholders abandon the land, secondary forests and scrub growth envelope the abandoned fields where fire has not destroyed seeds and seed sources. When smallholders sell their lands, they have usually worked only a small fraction of the land. The first generation's poverty and isolation limits the damage that they do to the forests, so, through a form of passive protection, a sizable proportion of the forests on populist frontiers continues to exist.

The pace of deforestation accelerates when the state subsidizes settlement by building roads and providing credit to smallholders. Most smallholders establish a claim to a tract of forested land with the hope that it will soon become accessible by road. To make their dream come true, smallholders form coalitions with land speculators to push, in concert with local politicians, for feeder road construction. These coalitions play a pivotal role in the extension of feeder roads into the countryside (Rudel with Horowitz, 1993). Once built, the feeder roads enhance access to urban markets and create incentives for smallholders to convert additional forests into fields.

When governments built feeder roads during the 1980s and 1990s, the incentives to convert forests into pastures were especially strong because expanding urban markets for beef in Amazon cities kept cattle prices high (Faminow, 1998). The increased access to cities also made mixed dairy–beef operations possible. Subsidized credit also accelerated land clearing. A bank loan provided the capital to purchase additional cattle.

To feed the larger herd, smallholders converted additional tracts of forest into pastures (Rudel with Horowitz, 1993). For wealthier smallholders, loans made it possible to intensify production through the introduction of improved pasture grasses, green manure, and fencing. These innovations made both beef and dairy operations more profitable and gave landowners incentives to expand their operations at the expense of the forest. In sum, when the state provided subsidies, smallholders overcame the limitations of infertile soils, established profitable enterprises, and converted almost all of their forested land into pastures (Vosti et al., 2001).[5]

For Brazilian smallholders with reliable access to markets during the 1990s, the economic returns from the labor of raising cattle exceeded by a ratio of 7 to 1 the returns from the labor of gathering forest products. The relative profitability of cattle ranching characterized farms with poor soils as well as farms with rich soils. Under these circumstances, only steep slopes and wetlands prevented smallholders from converting forests into pastures (Vosti, Witcover, and Carpentier, 2002:xi–xii). Government subsidies for cattle ranching declined but did not disappear during the 1990s. The profitability of cattle ranching persisted despite the decline in subsidies, because urban consumers used their increments in income to increase the amount of meat in their diets.

Even in this favored setting, not all smallholders experienced economic success. In some instances in Rondônia, impoverished smallholders in distress sold their lands to their more affluent neighbors, so landholdings began to concentrate in fewer hands along some populist frontiers. Some failed smallholders then took over other undeveloped farms and began clearing land. Poor rural residents continued to migrate to the forest margins because, without much education and with few opportunities to work in long-settled rural areas, they had few other choices. Spontaneous settlement at the margins of the forest was in effect a self-help land-redistribution program (Okoth-Ogendo, 1986:173).

During the 1990s, the number of potential migrants to the forest margins from both inside and outside the Amazon region declined from its peak in the 1980s (Perz, 2002), but the pace of the decline was irregular and variable from place to place. A complete record of poor migrant arrivals to Altamira shows a high rate of arrivals in the 1970s and early 1980s, followed by a decline in the early 1990s and a resurgence in the mid-1990s (Moran, Brondizio, and McCracken, 2002). Although the number of migrants to rural places has declined, the rates of deforestation have not followed suit because existing landholders have continued

> **Alfonso Caivinagua: Smallholder Persistence and Productivity**
>
> Born in the Andean highlands of southern Ecuador in the early 1920s, Alfonso Caivinagua first encountered the jungle in the late 1930s when stories of gold prompted him and thousands of other young *mestizo* men to make the arduous trek from the Andes to the Amazon and to begin panning for gold in the upper reaches of Amazon rivers. A few months later, a war with Peru and hostile encounters with lowland Amerindians *(Shuar)* persuaded Alfonso and his friends to return to the highlands. For the next three decades Alfonso tilled eroded soils in the drought-prone Andean highlands, but with little success. In the late 1960s, he joined an internationally financed, government-sponsored colonization project and received 50 hectares of Amazon rain forest for farming. A hard worker, Don Alfonso cleared almost all of this land during the 1970s and slowly built a herd of 40 to 50 cows. His children grew older, and one of them, Pepe, lost his own cattle to thieves in the early 1990s. In danger of losing his farm to the bank, Pepe decided to migrate to the United States illegally to work. To finance his son's trip, Don Alfonso sold his remaining cattle and took out a loan from a moneylender. The money from the loan paid for Pepe's trip to the United States. Now more than 80 years old, Don Alfonso still has productive pastures, and he has begun to build his herd once more. With his eyesight failing, he worries about not being able to see poisonous snakes in the pasture grasses when he works, but with no other economic alternatives, he soldiers on.
>
> SOURCE: Field notes, 2002.

to clear land for pasture at high rates. The story of Alfonso Caivinagua, an aging colonist in the Ecuadorian Amazon, illustrates the tenacity that many smallholders have exhibited in developing and then sustaining their cattle ranching enterprises (see box).

Corporate Frontiers: Mineral Deposits, Ecological Modernization, and Deforestation

Corporations built roads and railroads into places of extraction to get minerals out to world markets, and the new infrastructure shaped the local paths of rain forest destruction. The first companies to work in these places frequently exhibited a "take the riches and run" ethic, with no concern for the fate of the surrounding forests. In Ecuador, the first consortium of oil companies to work in the region built access roads along the pipelines, and colonists then used these roads to open up areas for settlement and deforestation, thereby magnifying the environmental impact of the oil extraction (Hiroaka and Yamamoto, 1980). By the late 1980s, a coalition of indigenous and environmental groups had begun pressuring the corporations to clean up their operations. Indigenous groups sued Texaco in American courts for oil-related damages to the rain forest environment.

Because the companies did not want to jeopardize their continued access to the oil field, they began an uncoordinated effort to limit the environmental damage caused by their operations. In the mid-1980s, to prevent indiscriminate deforestation of lands inside the Yasuni National Park where they were drilling for oil, Occidental Petroleum established gates on their access roads to the oil wells and did not allow colonists to use the roads, thereby reducing rates of deforestation near the oil wells. In the early 1990s, the Atlantic Richfield Company (ARCO) developed one sector of the oil field without building roads. Company employees used helicopters to transport drilling equipment and construction materials to the sites of the oil wells. To construct and maintain pipelines from the wells to the refinery 130 km to the east, ARCO engineers constructed a monorail that paralleled the pipeline. By using a monorail rather than a road to maintain the pipeline, the company reduced the damage to the surrounding forest. Typically, the construction of a road along a pipeline requires cutting a corridor at least 30 m wide through the rain forest; in comparison, the monorail–pipeline combination requires cutting only a 4-m-wide corridor through the forest (Alexander's Gas and Oil Connection, 1998). Most recently, companies have begun to try to capture and sell the natural gas emitted by the oil wells; this change would eliminate the air pollution from the flares at each well. Each of these innovations represents ecological modernization—the substitution of newer, environmentally less destructive technologies for older, more damaging technologies (Sonnenfeld and Mol, 2000).

Companies engaged in other mineral extraction enterprises elsewhere in the Amazon basin have undertaken similar initiatives. In an effort to prevent deforestation by colonists along company-constructed roads, Shell promised to restrict access to their natural gas concession in the Peruvian Amazon (Terborgh, 1999:54–55). As part of a joint effort with Conservation International in the Peruvian Amazon, Mobil reduced the land clearing that they do when, in searching for oil, they cut seismic lines through the rain forest (Thomsen et al., 2001:103–104). To avoid the expense of building roads to each drilling site, a gold mining company in

French Guiana has begun to use a small, portable drill that can be transported through forested land to drilling sites (Graybeal, 2001:228).

Companies make these changes under political pressure in an attempt to protect ongoing operations that continue to yield large profits. Viewed historically, the adoption of these new practices in mineral exploitation should reduce the amount of rain forest destroyed during the development of each new mine or oil field. In other words, the amount of forest destroyed during the development of the first block of oil fields in the Ecuadorian Amazon during the 1970s should clearly exceed the amount of forest destroyed during the development of similar-sized blocks during the 1990s. Despite these efforts at ecological modernization, each new mining operation does destroy additional tracts of rain forest.

Agribusiness Frontiers: Large-Scale Cattle Ranching, Soybean Cultivation, and Deforestation

Discouraged by the long distances to markets and the large amounts of capital required to turn frontier landholdings into profitable enterprises, many smallholders sell their landholdings to wealthier individuals intent on assembling large herds of cattle. If enough smallholders sell out, as they appeared to be doing around Sao Felix, Para, during the late 1990s (Maturana, 1999), then a populist frontier gradually becomes an agribusiness frontier, and the pattern of deforestation changes. Logging becomes part of the process in regions such as Para that have numerous mills and roads or rivers for transporting the logs to the mills. After the decline in state subsidies during the late 1980s, large landowners found a new source of revenue in the timber on their lands. In the 1990s, many of them logged their lands and used the profits from logging to convert the cutover areas into pastures for cattle.

Because pasture represents such an extensive use of land and requires little labor, ranchers seem insensitive to the poor quality of the soils beneath their pastures or the labor demands of additional tracts of pasture (Moran, Brondizio, and McCracken, 2002). They are sensitive to the availability of credit. During economic recessions, when banks and the government restrict the availability of credit, as occurred in the late 1980s and early 1990s, little land clearing occurs. When economic trends make credit more available, as occurred in the mid-1990s, deforestation rates increase (Moran et al., 2002). Ranchers sponsor political action groups who work to secure favorable terms for farm credit (Cowell, 1990: 137–148).

Soybean farmers in Brazil and Bolivia followed a similar pattern in clearing the semi-deciduous forests along the southern margins of the Amazon basin in Bolivia and Brazil (Hecht, 2002). Well-financed growers mechanized their operations and converted sizable tracts of forest into large fields for soybeans. Agronomists facilitated the agricultural expansion by creating new varieties of soybeans for these regions. Like cattle ranchers, Brazilian soybean growers organized politically and successfully sought government assistance when price and credit trends eroded the profitability of their enterprises during the late 1980s (Kaimowitz and Smith, 2001).

Price trends supported agricultural expansion. Given the relatively high income elasticity for beef products (greater than 0.6) in South America, prices for beef remained high throughout the 1980s and 1990s even though the volume of production from the Amazon basin soared (Faminow, 1998).[6] With an income elasticity similar to that of beef, soybeans prices also remained high during the 1990s (Kaimowitz and Smith, 2001). Under these circumstances, producers did not hesitate to increase production when credit became available, and in many instances, landowners expanded production by converting additional tracts of rain forest into fields. The only countervailing trend involved expanded production through intensification. Soybean farmers' heavy expenditures in preparing new soybean fields and the willingness of ranchers in old colonization zones to renovate pastures (Nepstad, Uhl, and Serrao, 1991) suggest that large landowners may be becoming more sedentarized and less willing to increase production by expanding the area under cultivation. This trend, if it becomes pronounced, could eventually lower deforestation rates in the basin.

Beyond the Margins of the Frontier: Reserves and Amerindian Preserves

Some regions show little change in expanse of forest that covers them, and, ironically, these places, far from markets and roads, often have legal protections as well. Although parks still constitute only about 3.0 percent of the land area, Amerindian reserves take up more than 15 percent of the land area in the Brazilian Amazon (Harcourt and Sayer, 1996:240–242; FUNAI, 2001). These lands cluster in the least accessible, "refuge" regions, in the northern and western portions of the basin, away from transportation corridors.[7] With few interest groups clamoring for the development of these regions, governments have found it politically feasible to designate these areas as parks or reserves for Amerindians. The conservationist ethics of park rangers and unacculturated indigenous peoples reinforces the passive protections provided by long distances to urban markets.

Conclusion

The spatial clustering of different land-use conversion processes has contributed to the emergence of regionally differentiated zones of land use within the basin, some places devoted to forests and other places devoted to pastures. By separating different types of land users, these schemes promise to build, over time, groups of users who have a stake in maintaining the prevailing land use in a region. For example, during the 2002 fire season in the Amazon, the forest-destroying fires frequently stopped at the edges of the Amerindian reserves because Amerindians patrolled the borders and fought fires when they found them (Rohter, 2002). This policy of spatial clustering would, for example, separate logging from cattle ranching by creating a network of large state forests that would permit logging and prohibit land clearing for pasture (Verissimo, Cochrane, and Souza, 2002; Schneider et al., 2000). Over time, logging firms would develop an interest in preserving state forests against incursions from ranchers because only this type of defense would preserve loggers' long-term supply of raw materials. Amerindian reserves that promote the extraction of nontimber forest products and the cultivation of perennials would also over time develop an indigenous constituency who would defend the reserve against incursions by outsiders.

In the frontier zones along the arc of deforestation, where most land clearing, logging, and fires have occurred, land-use planning by municipal and provincial authorities may produce some limited conservation gains (Nepstad et al., 2002), but, given the difficulty that local land-use planners have long had in resisting pressures from local interest groups to revise their plans to accommodate new development (Mahar and Ducrot, 1998), it seems unlikely that land-use planning could preserve appreciable amounts of rain forest in this region. Incentives to maintain woodlots on smallholdings in largely deforested regions may prove feasible (Vosti, Witcover, and Carpentier, 2002). Given that the Kyoto Protocol does not authorize payments under its Clean Development Mechanism (CDM) for preventing deforestation, the CDM would not seem to offer a means for preserving forests in these locales. Conservation easements through environmental nongovernmental organizations such as Conservation International may provide a means for preserving biodiversity in some of the remnant forests in frontier regions.

The political impetus for such large-scale resource partitioning usually emerges only in an atmosphere of crisis. For example, when the Brazilian government announced the record high rates of deforestation in 1995, essentially the result of the *terra firme* activities depicted in cells two and four of table 4.2, Fernando Enrique Cardoso, then President of Brazil, pledged to expand the protected area system depicted in cell five in table 4.2. It may take this type of dynamic to create the political opening necessary to launch the large-scale expansion of Brazilian state forests, protected areas, and Amerindian reserves that some analysts see as crucial for preserving a self-sustaining Amazon forest.

West Africa: From Cocoa Groves in Forests to Food Crops in Scrub Growth

At high-water you do not see the mangroves displaying their ankles in a way that shocked Captain Lugard. They look most respectable, their foliage rising densely in a wall irregularly striped here and there by the white line of an aerial root, coming straight down into the water straight as a plummet until it gets some two feet above water level and then spreading out into blunt fingers with which to dip into the water and grasp the mud. Banks at high water can hardly be said to exist, stretching away into the mangrove swamps for miles and miles.... At the ebb ... the foliage of the lower branches of the mangroves grows wet and muddy, until there is a great black band about three feet deep above the surface of the water in all directions; gradually a network of gray-white roots rises up, and below this again, gradually, a slope of smooth and lead brown slime.

MARY KINGSLEY, *Travels in West Africa* (1897)

The tropical forests of West Africa grow where humid ocean air masses bring rains to coastal portions of the region. The forests begin at the water's edge with mangroves south of Cape Verde and extend, with interruptions, along the coast in a narrow strip southward and eastward until

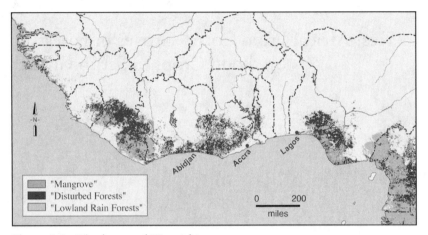

Figure 5.1   The forests of West Africa.

it merges with the large block of forest in central Africa (figure 5.1). One interruption occurs in Benin, in the Dahomey gap, where a mixed landscape of gallery forests and savannas reaches the sea. Despite their limited extent, the tropical forests of West Africa contain high levels of biodiversity. The Tai forest in Côte D'Ivoire, for example, contains at least 1,300 species of vascular plants, 700 of which are endemic to the region (Sayer, Harcourt, and Collins, 1992:26–27). The rains along the coast and a range of mountains, reaching 1,953 m above sea level and running parallel to the coast, have created an abundance of niches for plants.

The extent of the forests when Europeans first began to frequent the coast is the subject of some controversy, but by the late twentieth century, only small remnants of the sixteenth-century forests survived.[1] Shrub-dominated secondary growth with cultivated fields and small patches of forest currently characterize much of the West African landscape. The large extent of disturbed forest evident in figure 5.1 indicates the degree to which human activities have degraded the region's forests. In Nigeria, Ivory Coast, and Ghana, only inaccessible mangroves and forests in state reserves remain largely intact (Oates, 1999; Wagner and Cobbinah, 1993). Easy access explains the heavy human imprint on West African forests. The proximity of the forests to the coast attracted entrepreneurs from overseas with plans to extract riches from the region. More recently, with the development of an almost continuous belt of coastal cities from Douala in Cameroon to Abidjan in Ivory Coast, urban residents have increasingly made demands on forested regions for timber and foodstuffs.

As West African forests have declined in extent, they have become the world's most fragmented tropical rain forests (Rudel and Roper, 1997b). An extensive network of rural roads and tracks to agricultural holdings surround the remaining islands of forest and make them more accessible, so they are vulnerable to human interventions of all kinds. Historical trends in forest cover change are consistent with this dynamic. During the 1980s, regional forest cover declined by 0.8 percent per year; during the 1990s, the decline accelerated to 1.5 percent per year (FAO, 2001).[2]

West Africans are much poorer than people in Central and South America; they earned an average of $342 per capita in 1999. They are also increasing in numbers more quickly: population growth in the region averaged 2.7 percent per year between 1980 and 1999 (World Bank, 2001). Unlike the populations of both Central and South America, a majority of West Africans live in rural places, but rural to urban migration is rapidly redistributing the population toward cities. Between 1980 and 1999, the proportion of urban residents in West African societies increased from 28.6 percent to 43.3 percent (World Bank, 2001). Despite the stream of migrants from villages to cities, the absolute number of people in rural districts continued to increase during the 1980s and 1990s.

Paths of Destruction Across West African Forests
New Crops in West African Forests, 1500–1900

Most of the West African kingdoms during the Middle Ages occupied the savannas north of the West African forest belt, so the region's forests remained relatively intact at the beginning of the sixteenth century. Despite pillaging, robbery, and assault by slave traders, the number of shifting cultivators in the forests of Ghana probably increased during the eighteenth and nineteenth centuries. New food crops from the Americas increased the agricultural returns to shifting cultivation after 1510 A.D. Cassava, with its large volume of production, its slow rate of spoilage before harvest, and its ability to be stored in the ground for several years, contributed to the long-range food security of households. Maize, another transplant from the Americas, alleviated short-term food security with its short time to harvest. The simultaneous introduction of these two crops increased the returns to shifting cultivation, but they did nothing to alleviate the arduous labor of clearing primary rain forest with primitive tools, so when cultivators chose new locations for their fields, they preferred secondary forests. Over time, the extent of the secondary forests grew as populations of shifting cultivators grew. The agricultural surpluses from shifting cultivation, coupled with gold mining, provided crucial economic support for the Asante kingdom that emerged in the forest belt during the eighteenth and nineteenth centuries (Wilks, 1978; McCann, 1999:109–138). By the beginning of the twentieth century the practice of shifting cultivation had significantly reduced the extent of old growth forests in West Africa.

Settlement in the rain forest region remained somewhat uneven. Higher-elevation locations reduced people's exposure to malaria, so settlements of groups such as the Krobo, the Shai, and the Akwapim clustered in the hills and along the ridges of the Gold Coast. The lowlands contained large stretches of less densely populated old growth forests. With the beginning of the cocoa trade in the late nineteenth century, the spatial distribution of the population changed (Hill, 1963:25–27).

1900–1985: Cocoa Frontiers

The coastal location of West African forests close to Europe encouraged early traders to begin exporting the products of tropical agriculture to Europe. The international trade in cocoa developed toward the end of the nineteenth century, and smallholders moved quickly to establish cocoa plantations. In Ghana, smallholders abandoned highland villages and staked out claims for cocoa farms in lowland jungles recently made more accessible by the extension of railroads inland from ports on the ocean. Smallholders saw the conversion of forests into cocoa plantations as an investment. In the words of one early ethnographer, "The essential nature of the migratory process is that it is forward looking, prospective, provident, prudential—the opposite of hand to mouth. They did not 'eat' the proceeds from the early farms. Rather almost from the beginning the farmers regarded themselves as involved in an expansionary process from which they had no intentions of withdrawing. Almost from the beginning cocoa farms established on purchased land were regarded as investments—i.e., property which existed for the purpose of giving rise to further property" (Hill, 1963:180).

Anxious about a projected scarcity of timber from the destruction of the forests, foresters from British and French colonial services established

Table 5.1

Tropical Deforestation in West Africa: A Qualitative Comparative Analysis

A. Deforestation During the 1970s and 1980s (12/12)
  topo POP (3) +
  INTLMKT POP LOGGING (1) +
  topo intlmkt LOGGING (2) +
  POP intlmkt logging (1) +
  topo INTLMKT logging (2)

B. Deforestation During the 1990s (10/10)
  FORSIZE SMALLAG FIRE (1) +
  URBMKT FORSIZE SMALLAG (4) +
  urbmkt FIRE SMALLAG (2) +
  URBMKT smallag forsize fire (1)

The "+" in this table indicates that deforestation occurs whenever any of the listed combinations of causal conditions exists.

Upper case indicates that the causal factor is present, and lower case indicates its absence. The ratio in parentheses indicates the proportion of studies for that period whose findings agree with the Boolean expressions in the lines below it. The number in parentheses after each causal combination indicates the number of studies whose findings fit that causal conjuncture.

Factors: FIRE = frequent burning of fields; FORSIZE = forests cover greater than 40% of the local landscape; INTLMKT = region produces crops for export; LOGGING = timber exploitation, usually for export; POP = rural population increase; SMALLAG = smallholder agriculture; TOPO = mountainous terrain; URBMKT = expanding cultivation of food crops for urban markets.

networks of forest reserves in the early twentieth century. The creation of the reserves met with local opposition from forest farmers, who resented the restrictions on their activities. The designation of lands as a forest reserve did not, however, prevent either migrants or local residents from occupying lands in the reserves after independence (Martin, 1991:21). Beginning in the 1960s, the migrants to the forested, cocoa-producing zones came from the drier savanna regions to the north, oftentimes from other countries such as Burkina Faso. In some instances, as in Ivory Coast under Houphouet-Biogny, the government encouraged prospective migrants to carve farms out of the forest in the cocoa-producing zones (Ruf, 2001:297). The migrants encountered native residents in the forest zones, and, when the natives tried to prevent the migrants from planting perennials that would strengthen the migrants' claims to the land, the migrants occupied lands far from native villages. In reaction, native residents would begin clearing land in the buffer zone between the two settlements in order to establish their claim to these lands, so that they could sell the cleared lands to later waves of migrants (Martin, 1991:21–22). Under these circumstances, local rates of deforestation accelerated.

The case studies from the 1970s and 1980s in part A of table 5.1 chart the changes in a West African landscape that still contained appreciable amounts of forest and millions of small farmers. In places where loggers extracted timber from old growth forests (LOGGING), where smallholders cultivated cacao for export (INTLMKT), and where rural populations grew rapidly (POP), deforestation occurred. Logger-led deforestation began after World War II, when European investors noted the proximity of West African forests to the sea and overseas markets. They then obtained and logged forest concessions in Cameroon, Ghana, and Ivory Coast. The loggers' activities facilitated agricultural expansion by ever-larger populations of smallholders. By creating rudimentary roads in the forest to extract timber, the loggers made the cutover lands along the logging tracts accessible to outside markets. "Follow on" farmers took advantage of the ease of access and the partially cleared cutover lands to start cocoa farms.

A small farm along a logging road in Cameroon typically contained a house garden, one or two hectares of cocoa, bananas grown under a canopy of fruit trees, and plots of food crops such as rice, corn, or cassava (Barraclough and Ghimire, 2000:88). The land often belonged to a community, but individuals had rights to the fruit trees and perennials such as coffee and cocoa that they had planted (Leach, 1994:108–11). Because smallholders did as well as or better than large landholders in cultivating cocoa and coffee, landholdings in West Africa, unlike those in Latin America, tended to be broadly distributed (Kotto-Same et al., 2000:34).

The mix of products coming from the farm and the labor-intensive nature of both the field and tree crops created constant labor supply problems for farmers. Would a household have enough labor to harvest a rice plot and a cocoa grove during the same month? Men and women solved these problems both within and between households through a complicated and ever-changing set of labor-sharing arrangements (Leach, 1994:68–99). Under these circumstances, a large number of children solved a farmer's labor supply problems on his farm and made it possible for him to undertake larger-scale and sometimes more profitable endeavors such as the cultivation of oil palm. Temporary labor migrants from northern Ivory Coast and Burkina Faso also worked for forest zone farmers at harvest time (DePlaen, 2001). The qualitative comparative analysis outlined in table 5.1 testifies to the importance of rural population growth during the 1970s and 1980s in driving deforestation in West Africa. It produced prosperity in households and deforestation in landscapes.

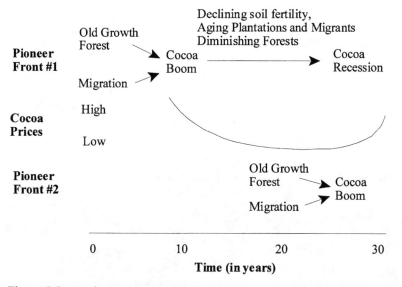

Figure 5.2 Cycles in cocoa economies and changes in deforestation fronts.

Farmers preferred to clear secondary forests rather than primary forests for cultivation because clearing secondary forests required much less work (Leach, 1994:146). The secondary growth around newly established villages changed with repeated use of the land. When farmers let fields go fallow for the first time, fast-growing umbrella trees *(Musanga cecropioides)* sprang up to fill holes in the canopy and provided shade for the seedlings of the tree species to follow, so secondary forests soon covered the site. The sequence changed if a farmer returned to the site several years later and destroyed the umbrella trees to cultivate the land. After this farmer left and the land was allowed to go fallow again, umbrella trees would no longer grow there, and forest would not reoccupy the site. Without the shade provided by the umbrella tree, tree seedlings did not germinate in the site, and a low "bush fallow" or "farm-bush" gradually spread over the site (Martin, 1991:132). With limited amounts of vegetation, the soil fertility of these lands recuperated very slowly, so many farmers abandoned these lands and looked for more fertile lands covered with primary or old growth secondary forests to farm.

Other changes, outlined schematically in figure 5.2, occurred during the boom times of rapid settlement on a cocoa-growing frontier, and these changes laid the groundwork for later economic distress. Harvests

Figure 5.3   A burnt secondary forest prior to planting, Guinea. Photo credit:
FAO photo (1984).

increased with the rapid growth in the amount of cultivated cocoa, so
prices for the commodity began to decline. Farmers gradually exhausted
the nutrients and organic matter that had accumulated in the soils while
the land was in forest (referred to as "forest rent" by economists). Plant
diseases that spread across the cocoa groves exacerbated the farmers'
production problems. The colonial regimes intervened to fight the diseases, but sometimes the cures seemed worse than the diseases. To avert
further damage from swollen shoot disease, British colonial administrators cut down thousands of smallholders' diseased trees during the 1930s
(Berry, 1993:51).

Uncultivated forest lands in the region declined in extent as other
farmers converted forests into cocoa groves, so farmers could not find
fertile forest lands nearby. Finally, the pioneering farmers grew older and
less willing to engage in the heavy work of felling trees in a primary forest. They were more likely to return to their place of origin and leave
their now less-productive farm in a tenant's or caretaker's hands (Dorm-
Adzobu, 1974). These trends together caused production to decline along
the older cocoa frontiers. The decline in production eventually induced
a rise in the price of cocoa. In response to the price rise, young migrants

opened up a new cocoa frontier elsewhere in the region or in Southeast Asia, and the cocoa cycle spawned a second wave of deforestation (Ruf, 2001).

The sequence of events just outlined recurred in different parts of the forested belt of land that runs along the coast of West Africa. A different dynamic characterized the transition zone to the north, where savannas cover most of the land and gallery forests constitute only 10 to 15 percent of the land area. The scarcity of wood in this region induced smallholders to afforest—to plant fruit-bearing trees and other commercially useful trees on savannas in close proximity to their homesteads. Differences in fire dynamics and in the land-use practices of farmers on lands farther from their homes created a varied pattern in land use in this region. In some regions, such as Kissidougou prefecture in Guinea, net gains in forest cover occurred (Fairhead and Leach, 1996:55–85); in other regions, such as Kilimi, Sierra Leone, net losses in forest cover occurred (Nyerges and Green, 2000). The historical record of land-cover changes in these forest-scarce settings suggests that when supplies of labor were abundant, farmers used some of that labor to create more forests.

1985–2000: Urbanization, Food for Urban Markets, and Forests

The qualitative comparative analysis of 22 case studies of deforestation in West Africa during the 1980s and 1990s, presented in table 5.1, suggests the changing dynamics of landscape change in the region during the last two decades of the century. Part B of table 5.1 presents the causal configurations for deforestation in the 1990s. Land clearing in the 1970s and 1980s changed the landscape and altered the dynamics of deforestation during the 1990s. It concentrated in the small number of districts that still contained large stands of forest (FORSIZE). Local urban markets for foodstuffs (URBMKT) provided income for most farmers. Finally, by destroying so much forest, the landscape transformations of the preceding decades had dried out the landscape, causing fires (FIRE) to occur more frequently. The fires destroyed seed sources and prevented forests from regenerating in some places. In the aftermath of the fires, an invasive, fire-resistant South American shrub *(Chromolaena odorata)* occupied extensive tracts of land along the old cocoa frontiers.

Changes in macroeconomic conditions during the 1980s altered the regional landscape. During the 1970s and early 1980s, government mar-

---

Michael Kwabla Odjidja: What Does an Ex–Cocoa Grower Do?

Mr. Odjidja was born around 1930 and he grew up in the Krobo district, a cocoa-growing region about 90 km north of Accra, the capital and largest city in Ghana. His father practiced traditional medicine, and from him Odjidja learned about the properties of plants in the region. He first found work in nearby diamond mines, but after World War II he began to work for the state's cocoa cultivation extension service. Farmers began establishing cocoa groves during the first three decades of the century in Krobo, but by the 1930s, fires, drought, and swollen shoot disease had begun to afflict cocoa groves in the district, so the state established an extension service to assist growers. For many years, Odjidja maintained a cocoa tree nursery for the extension services. Through this work and knowledge acquired as a child, Odjidja became very knowledgeable about plants and agriculture in the district. After retirement from the cocoa service, he assisted Kojo Amanor in a study of how farmers in Krobo coped with environmental degradation in the region. To provide an income for himself during his declining years, Odjidja started a nursery for nyabatso *(Newbouldia laevis)*, a pioneering tree in secondary forests that has medicinal properties. He hoped to sell seedlings to other smallholders. In his long journey from medicinal plants in his childhood to cocoa in adulthood and then to nyabatso during retirement, Odjidja recapitulated the journey that many small farmers in West Africa made during the twentieth century, from local collecting to export-oriented agriculture and then to a more locally oriented agriculture. Odjidja died around the turn of the century.

SOURCE: Amanor (1994:x, 80, 213–214).

---

keting boards usually purchased cash crops such as cocoa from farmers and subsidized the purchase of inputs such as fertilizers. With the onset of the debt crisis in the mid-1980s, every state in the region had to negotiate an agreement with the International Monetary Fund (IMF) to obtain emergency loans. To obtain these structural adjustment loans, states had to cut expenses, eliminate market distortions, and generate more revenues. Unable to generate much additional revenue, governments lived from loan to loan.

In Ghana, a cash-starved government decided to increase revenues from the export of logs through a generous program of credits and concessions to logging firms (Owusu, 1998), so timber companies continued to contribute to deforestation during the 1990s, just as they did after World War II, by building roads to extract timber from the forests. Each

IMF-recommended devaluation expanded the overseas markets for West African timber.

Regimes also had to make politically painful cuts in services to rural people. To cut expenses, public officials in Cameroon ended price controls on cocoa, eliminated subsidies for agricultural inputs, abolished government marketing boards, and began to promote export-oriented industries in an effort to increase government revenues from exported goods. The loss of state subsidies for agricultural inputs made it economically impossible for many smallholders to continue to cultivate cocoa. (For the experience of an ex-cocoa grower in Ghana, see the box.) Smallholders who persisted in cultivating cacao without subsidies began to replant old cocoa lands, because primary forests with fertile soils no longer existed in many parts of the region (Benhin and Barbier, 2001). Here and in Ghana, cocoa frontiers receded in importance as a cause for deforestation during the 1990s.

After the loss of subsidies made cocoa cultivation less profitable, many farmers in Cameroon opted to cultivate food crops such as cassava, plantains, cocoyams, and vegetables for the growing urban markets. Unlike coffee and cocoa, which grow under a shade canopy, the food crops grow in open fields, so land cleared for these crops undergoes complete deforestation as opposed to the partial deforestation that occurs when smallholders convert forests into cocoa fields. Although the cash income from the food crops in Cameroon did not compare with the returns from cocoa ($644 versus $1,755), it secured a smallholder's subsistence and provided an alternative source of income (Kotto-Same et al., 2000). Because subsistence concerns shaped many household decisions to clear land during this new period of austerity, population growth among smallholders compelled agricultural expansion (Wunder, 2003). In this endeavor, access to land mattered more than access to markets, so smallholders sometimes cut fields out of forests far from roads (Mertens and Lambin, 2000).

In older colonization zones in Ghana, farmers abandoned cocoa groves because they could no longer compete with growers in newer production zones. The smallholders who remained in the old zones turned to foodstuffs such as cassava and fruit-bearing trees as sources of income. Traders traveled the dilapidated roads of these districts, buying fruits and other foodstuffs for resale in the markets of large coastal cities such as Accra or Abidjan. Many of the young have left these zones for the cities, and others have found nonfarm sources of income in the region. Since the early 1980s, fires have coursed across these predominantly agricultural landscapes, and C. *odorata* has occupied extensive areas (Amanor, 1994).

Although most West African states withdrew vital services from rural areas in response to IMF financial strictures, some states stopped providing all services outside of major cities. The collapse of these states and the ensuing political chaos affected the forests in geographically uneven ways. Political turmoil engulfed Liberia and Sierra Leone in the late 1980s and early 1990s, endangering old growth forests and encouraging secondary growth. The collapse of central authority created open-access conditions in the hinterland, so that anyone with political and economic resources could appropriate natural resources such as diamonds or commercially valuable timber in old growth forests. The warring parties had desperate needs for resources, so they encouraged their collaborators to log accessible stands of timber (Ellis, 1998; Abdullah and Muana, 1998). The sharp increase in reported deforestation rates between the 1980s and the 1990s, from 0.5 to 2.0 percent in Liberia and from 0.6 to 2.9 percent in Sierra Leone, reflects this assault on old growth forests. In rural places from which people fled for their safety, secondary and scrub growth spread over the abandoned fields. In places where refugees sought sanctuary, their arrival put a strain on local rules for regulating natural resource use. For example, when large numbers of Liberians sought refuge in Sierra Leone in the late 1980s, they began chopping down palm trees to eat the fruit despite rules against this practice established by the elders of the Sierra Leone villages around the refugee camps (Leach, 1994:155).

Trajectories of Change in West African Landscapes

For many conservationists, West Africa presents an image of a future that all rain forest regions may soon face. Forests exist only inside the boundaries of reserves surrounded by poor and hungry migrant farmers who, in this pessimistic vision, gradually encroach on the remaining tracts of rain forests within the reserves until the forests are gone. In the late 1980s, conservationists initiated integrated conservation and development projects (ICDPs) with the hope that a mix of economic opportunities and conservation initiatives in the immediate vicinity of the parks would prevent growing rural populations from exploiting the few remaining rain forests within the parks. Later reports by conservationists have concluded that ICDPs were not effective (Oates, 1999; Terborgh, 1999). These findings are not surprising, given that the communities around the parks, which usually have to enforce the conservation measures in ICDPs, have short histories. In many instances, migrants founded the communities only a few years earlier, and under these circumstances the elders of the communities have little authority over other residents. Given the ineffectiveness of ICDPs, conservationists have begun to advocate coercive conservation, an increase in the number of armed guards around the reserves, as the only way to ensure the continued survival of rain forest flora and fauna in West Africa (Terborgh, 1999).

This prescription for rain forest conservation seems incomplete.[3] In particular it ignores both the ecological and economic potential of the extensive tracts of secondary vegetation that now exist in the West African rain forest belt. When the large islands off the West African coast, Fernando Po, Sao Tome, and Principe, became independent during the 1960s, growers with colonial ties abandoned extensive tracts of tired cocoa lands. The islands now contain more secondary forests than primary forests (Fa, 1991; Jones, Burlison, and Tye, 1991). On the mainland, secondary vegetation predominates in the formerly forested regions of countries such as Ivory Coast (Martin, 1991:132). Smallholders in Ivory Coast and elsewhere have gradually begun to learn how to exploit the agronomic potential of these old fallows. Farmers found that they could achieve higher rates of seedling survival if they grew cocoa seedlings in plastic bags in nurseries and then transplanted them into fallows. Although smallholders have not established timber plantations on bush fallow land, they have begun interplanting timber trees with cocoa on old fallows as alternative sources of timber have declined and prices for it have risen.[4] In some old fallows, smallholders found that hoe plowing and interplanting cocoa with yams and plaintains produced cocoa plants that were more disease resistant (Ruf, 2001:301–302).[5] Because these rehabilitated cocoa groves always grow beneath a canopy of larger trees, rehabilitating old cocoa lands almost always involves reforesting and, most likely, will increase the biodiversity of these lands.

Two other related trends have encouraged smallholders to rehabilitate degraded lands. First, by the late 1990s, longtime residents had begun to resist migrant attempts to occupy the few remaining areas of unclaimed land in Ivory Coast (Bassett, 2001). Interethnic conflict between northern migrants and southern residents intensified after 2000. Civil war broke out between the north and the south in Ivory Coast in 2003. The increasing levels of violence between migrants and locals has curtailed migration to the more forested zones and encouraged the younger generations in migrant families to become more sedentary, secure their tenure in the land, and to invest in old fallows.[6] Second, continued rural population growth puts more pressure on the remaining primary forests, but it also ensures that smallholders will continue to have the household labor necessary to rehabilitate farm-bush. In this sense, continued rural population growth creates a path dependency, pushing West Africans farther down a path of more intensive smallholder cultivation that, relative to other regions of the tropical world, they have already chosen.

These smallholder efforts seem likely to become more prevalent because the rapid growth of the coastal cities has created large markets for locally grown foodstuffs throughout the region. Between 1970 and 1990, the percentage of regional agricultural income generated by local urban markets increased from 56 to 73 percent (Cour and Snrech, 1998:51). In response to these incentives from urban consumers, smallholders have begun to afforest agricultural landscapes with a mix of trees that provide either forest or commercially valuable timber. Fairhead and Leach's (1996) work on smallholder afforestation efforts in the mixed forest–savanna zone in Guinea represents a somewhat different, but still recognizable, agro-ecological precursor to the afforestation efforts described here.

West African governments have begun to encourage community afforestation efforts through forest law reforms. Since 1990, governments in eight West African countries have passed forest reform laws that have given communities more control over the management of nearby forests and a larger share of the proceeds from the sale of forest products from these forests. These reforms, coupled with sustainable development projects, have provided villagers with incentives to initiate locally led afforestation efforts, either on their own lands or on those of communities.[7] By 2002, villagers in the Gambia had established 233 community forests (Wily, 2002). Aggregated, these efforts on private and communal lands sometimes have a significant effect. During the 1990s, the Gambia was the only West African country to experience an increase in forest cover (FAO, 2001). Given the difficulties experienced in establishing the community forests, most of the increases in forest cover must have occurred in woodlots on small farms (FAO, 2003a).

Other, as yet unadopted, institutional changes would intensify smallholder efforts to rehabilitate the West African landscape. Intensive cocoa cultivation, intercropped with timber or fruit trees, responds well to agrochemical inputs, but smallholders will only use them if rural markets, credit programs, and technical assistance become routinely accessible. This prospect would be enhanced if the World Bank's program of debt relief for heavily indebted countries allows West African countries to resume road building and agricultural services in long-settled rural areas.[8] If producers can establish connections to markets overseas, they may decide to make a virtue out of the necessity for low-input agriculture imposed on them by poverty. Given their ready access to European markets with large numbers of consumers concerned about the effect of chemicals on the food supply, West African smallholders could produce certified organic chocolate and coffee.

The trends in smallholder use of bush fallows, along with the institutional changes, suggest a strategy for preserving tropical forests that is less focused on the remaining stands of primary forest in parks. By regarding farm-bush as a significant resource that, through rehabilitation, can shelter some biodiversity and provide smallholders with some income, this strategy attempts to take smallholder pressure off the remaining primary forests. If successful, this strategy would not do away with the need for park guards. Guards and researchers, financed from outside the region, may provide the only effective, short-term means for limiting hunting inside the reserves (Oates, 1999:60). Like most sustainable development efforts, this plan is a hybrid. It tries to conserve biodiversity while at the same time delivering agricultural benefits to smallholders who have already begun to rehabilitate degraded bush fallow lands.

6

Central Africa: Passive Protections for Rain Forests

The Father [a cataract below Kinshasa on the Congo] is the wildest stretch of river I have ever seen.... Some of the troughs were a hundred yards in length, and from one to the other the mad river plunged. There was first a rush down into the bottom of an immense trough, and then, by its sheer force, the enormous volume would lift itself upward steeply until, gathering itself into a ridge, it suddenly hurled itself 20 or 30 feet straight upward, before rolling down into another trough. If I looked up or down along this angry scene, every interval of 50 or 100 yards of it was marked by wave towers—their collapse into foam and spray, the mad clash of watery hills, bounding mounds and heaving billows, while the base of either bank, consisting of a long line of piled boulders of massive size, was buried in the tempestuous surf. The roar was tremendous and deafening.
H. M. STANLEY, *Exploration Diaries* (1961)

The Father, described by Henry Morton Stanley after his descent of the Congo, is only one of 33 cataracts, spread out over 230 miles just before the Congo meets the sea. Explorers christened the last of these rapids "the Cauldron of Hell." The river that rushes through the Cauldron of Hell drains the eastern two thirds of the central African rain forest. For the past 500 years, outsiders have tried to exploit this forest, but the rapids have wrecked their schemes. The cataracts make it impossible to move goods in and out of the basin by boat, so they have to be transferred from boats to roads or railways to bypass the rapids. The transshipment of goods has added extra costs and made it difficult for farmers and loggers to sell products to the outside world at competitive prices. The climate and soils of the basin above the rapids provide additional challenges to people who would destroy the forests to expand agriculture. Large swamps and infertile soils, leached by heavy rains, have deterred people from exploiting forested lands in the Cuvette Central of the Congo River basin (Barnes and Lahm, 1997).

The central African forest extends for 1,300 miles west to east, from the Atlantic Ocean, across a low ridge of hills just inland from the coast, to the highland lake and volcano region along the Albertine rift in the center of the continent. The forest extends 700 to 800 miles north to south, from the savanna–woodland mosaics that border the Sahara to the open, miombo woodlands of southern and eastern Africa. It is the second largest contiguous block of tropical rain forest in the world (FAO, 2001:121). The Democratic Republic of the Congo (DRC), Congo–Brazzaville, and Gabon contain most of the forest, but it also extends into southeastern Cameroon, the southernmost regions in the Central African Republic, the continental portions of Equatorial Guinea, and northern Angola (figure 6.1).

The central African forest is the least threatened of the world's rain forests. The Food and Agricultural Organization of the United Nations (FAO) reported a slow rate of decline in the region's forests during the 1980s, 0.5 percent per annum. In the 1990s, the forests declined in size at a still slower rate, 0.3 percent per annum (FAO, 1993, 2001). Even at these slow rates, the FAO may be overestimating the amount of deforestation occurring in the region. A detailed analysis of estimates of land-cover change for Gabon by Sven Wunder (2003:84–90) suggests that, contrary to the FAO analysis, forest cover increased rather than decreased between 1970 and 1990. The contrast with forest cover trends elsewhere raises an obvious question. Why have the central African forests survived when the forests in so many other tropical regions have disappeared?

Impassable rapids, infertile soils, and swamps have long provided passive protections for central African forests, but in recent decades human

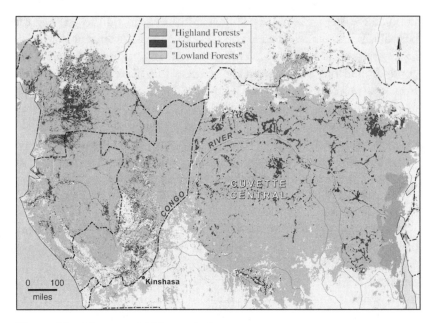

Figure 6.1   The central African rain forest.

affairs have become steadily more important in shaping land-use dynamics. Booms in mineral exports, coupled with political turmoil, have accelerated rates of urbanization that in turn have contributed to a pattern of peri-urban deforestation and interior afforestation in central Africa. After discussing the impact of colonial regimes on tropical forest cover, I will provide a more detailed explanation about the ways in which these human forces have shaped recent forest cover change in central Africa.

Forests and Colonial Regimes, 1880–1975

When French, Belgian, German, and Portuguese colonizers imposed their authority on the peoples of central Africa between 1890 and 1920, they encountered small, widely dispersed populations. For the previous three centuries, traders had gradually incorporated most of the basin's people into a global network for exchanging goods. First, traders sent slaves and ivory downstream; in return, they received European muskets, cloth, and alcohol. Cassava also became a commonly traded commodity as cultivators began to supply foodstuffs for the growing populations around the

Pool, the settlement that would later become Kinshasa, just above the rapids on the Congo River (Harms, 1981:1–71). The increase in trade did not benefit many of the basin's people. For two centuries, slave traders systematically killed villagers who resisted capture and sale. Imported diseases exacted a fearsome toll in lives. Epidemics of smallpox and sleeping sickness killed hundreds of people in the small villages along the main stem of the Congo River during the first two decades of the twentieth century (Harms, 1981:231–232). Already small, the human imprint on the central African forests declined even more in the aftermath of the epidemics.

To make their colonies self-supporting, the colonizers tried to establish revenue-generating agricultural activities early in the twentieth century. The Germans in Cameroon, the French in Equatorial Africa, and the Belgians in the Congo all imposed coercive labor systems to make the small and widely scattered indigenous peoples work in the fields.[1] The Europeans concentrated Africans along roads and rivers, assigned them production quotas of commercially valuable crops such as cotton, and punished them if they did not meet the quotas (Likaka, 1997; Adams and McShane, 1992). In the Belgian Congo, the authorities forced peasants near the mines in Katanga to grow food for the miners. To facilitate the marketing of cotton in the Belgian Congo, administrators ordered the construction of "cotton roads" in the producing regions and forced some peasants to relocate their households to places along the road. District administrators assigned quotas of cotton production to the relocated peasant households. Colonial authorities tried to enroll almost the entire populations of some districts in these schemes. By 1927, 80 percent of the Uele population in the northern production region had begun to cultivate cotton (Likaka, 1997:21). Between 1917 and 1952, cotton acreage in the Belgian Congo increased from 112 hectares to 368,000 hectares, and cotton cultivators increased from a few dozen to more than 800,000 households (Likaka, 1997:42). Through this program, shifting cultivation in widely dispersed locations gradually gave way to corridors and spots of permanent agriculture along roads and around towns, missions, and mines.

The absence of systematic data on the extent of the forests makes it impossible to estimate the impact of colonial agriculture on the central African forests, but the small size of the regional population almost certainly limited the extent of the clearing to a small fraction of the forest. The coercive methods of the colonial authorities reflect in part their continuing difficulty in finding enough workers for their agricultural enterprises. Throughout the colonial era and especially after 1930, labor scarcity limited agricultural expansion. The impact of colonial agricultural expansion on the forests of the Belgian Congo might have been greater had colonial officials not encountered labor constraints to expanded cotton cultivation.

Africans resisted the oppressive cotton cultivation schemes by fleeing into the forest and forming fugitive communities, particularly in the northwestern Congo, far from colonial centers of power (Likaka, 1997:120–130). When mining began, the labor scarcities grew more acute because colonial officials wanted workers for copper mines in Katanga and diamond mines in Kasai as well as for cotton fields in the northern provinces. When the British began developing the "copper belt" mines in Northern Rhodesia, the flow of immigrant laborers to the mines in Katanga stopped, and the Belgians' problems of labor scarcity grew even more severe (Likaka, 1997:26). These indirect effects of mining on land-cover change became more pronounced after independence when major discoveries of oil transformed the economies of central African nations. In all of these instances, a scarcity of labor limited agricultural expansion and prevented the destruction of forests.

Forests in the Postcolonial Period, 1975–2000

Despite the mineral wealth, central Africans have not prospered during the postcolonial period. The economic booms accelerated urbanization, but incomes declined during the subsequent busts. Political unrest also impoverished people. During the 1990s, only two nations in central Africa experienced increases in per capita income. Throughout the region, incomes were very unequally distributed among households (FAO, 2003b). In central Africa, more than any other region on earth, poor people live in rich forests. This seeming paradox makes sense only when the economic effects of mineral resources become clear.

Oil Booms and Dutch Disease

For different historical periods during the twentieth century, the macroeconomics of the predominantly mineral economies in the region have indirectly protected the forests through a condition that analysts call Dutch disease.[2] It occurs when one sector of an economy "booms" and other sectors of the economy do not. Conditions in the boom sector attract labor from the nonboom sector and create a set of associated conditions that make it difficult for enterprises in the nonboom sectors to survive. Almost every central African economy has developed a large mining sector, and these economies depend on the foreign exchange generated by exports from this sector. In the late 1990s, oil exports made up 90 percent of the value of exports in Angola, 75 percent in Gabon, and 50 percent in the Republic of the Congo. Taxes on diamond and copper mining enterprises provide substantial amounts of revenue to governments in the Central African Republic and the Democratic Republic of the Congo (Central Intelligence Agency, 2001). Booms in the prices of these commodities have made these economies susceptible to cases of Dutch disease in which booms in the minerals sector weaken agricultural enterprises, prevent agricultural expansion, and preserve the forests (Wunder, 2003).

The oil booms began during the 1970s when energy companies began to exploit oil fields off the shores of northern Angola (Cabinda), Gabon, the Republic of the Congo, and, most recently, Equatorial Guinea. Because the companies drill for oil off-shore, their activities have had few direct impacts on landscapes in central Africa, but they have had substantial indirect impacts through the economic booms that they triggered in coastal nations. Large flows of revenue from the oil companies into the coffers of the central governments stimulated construction booms and the hiring of large cadres of government employees in the cities. In response to the creation of new jobs in urban areas, rural–urban migration accelerated. By 1999, an average of 49.4 percent of the populations in central Africa lived in urban areas compared with 30.5 percent of the populations in sub-Saharan nations outside of the region (World Bank, 2001). The population of Libreville, the capital of Gabon, almost doubled every decade, growing from 12,500 persons in 1950 to 338,000 persons in 1998 (Wunder, 2003:118). Brazzaville, the capital of the Republic of the Congo, grew from 455,000 persons in 1980 to 1,206,000 persons in 2000 (World Bank, 2001).

The booms also caused the real exchange rate for Gabon's and Cameroon's currencies to appreciate, making it difficult for coffee and cacao exporters to sell their products in international markets (Wunder, 2003). The same trends in exchange rates made it cheaper to import cereals from abroad, and urban consumers' tastes changed accordingly. In Gabon, people reduced their consumption of locally grown cassava and began to consume large amounts of imported grains and beef (Wunder, 2003:99–

100). Both the changes in exchange rates and the changes in consumer tastes made it difficult for local farmers to sell their products and, in so doing, discouraged the conversion of forests into fields.

A somewhat similar sequence of events unfolded in the newly independent state of Zaire. The new government depended on revenues from mineral exports (diamonds and copper) from the interior provinces of Kasai and Katanga, far from the capital, Kinshasa, where government officials spent much of the revenue from the mines. Under these circumstances, secessionist movements sprang up in the interior right after independence as locals tried to wrest control of the mineral revenues from the central government. With mining revenues flowing through the capital city, Kinshasa grew rapidly, from 2.217 million people in 1980 to 4.917 million people in 2000 (World Bank, 2001).

Peri-urban Deforestation

The urban landscape of Kinshasa looks green. From the upper floor of Kinshasa's tallest building the eyes pick up "a pattern of neat squares carved into the red earth. For kilometres around, all open spaces had been divided into carefully watered plots. On road verges, traffic islands, what should have been the lawns of the ministry itself, the distinctive spiky leaves of the cassava leaves grew. This was a green city, but it was not greenery aimed at pleasing the eye. While Mobutu amused himself landscape gardening in Gbadolite, a nation on permanent Low Batt (subsistence) had no time for lawns. Preoccupied with the immediate problem of getting enough to eat, the residents had turned Kinshasa into one massive vegetable allotment" (Wrong, 2001:137).

While urban residents try to meet domestic needs by cultivating gardens on small patches of land near their homes, small farmers on the outskirts of the cities grow food crops for urban markets. As the cities have grown in size, the extent of land devoted to intensive agriculture in peri-urban regions has increased. Even in Gabon, where Dutch disease has caused agriculture to languish, the area devoted to intensive agriculture outside of the three major cities probably increased by 60,000 hectares between 1950 and 1990 (Wunder, 2003:89). The concentrations of poor, urban residents have also created large markets for fuelwood and lightweight poles for construction. Road building in peri-urban zones has accelerated deforestation (Krutilla, Hyde, and Barnes, 1995). The new

---

Mama Kagu: Smallholder and Single Mother

Mama Kagu has spent her entire life along the northern edge of the Guineo–Congolian rain forest in communities of farmers that have begun growing foodstuffs for urban markets in Bangui, Brazzaville, and Kinshasa. As a young woman, she had children, but then she separated from her husband and returned to her mother's home. When her mother died, Mama Kagu and her children moved to Bogofo, where missionaries, after reopening a hospital in the 1980s, had encouraged people to establish homes. For a small price, Mama Kagu purchased several tracts of secondary forest in Bogofo, which she has since cleared and planted with peanuts. The emergence of a market in land has occurred during a period in which people have cleared considerable amounts of forest and reduced local populations of game through hunting. Mama Kagu attributes the decline in game to the rapid growth of the human population in the area. She uses a digging stick—a simple, labor-intensive technique to plant peanuts. With the proceeds from the sale of her peanuts, Kagu clothes her children and pays for their schooling. They feed her when she returns from the fields after a long day of work. She has no interest in remarrying and takes pride in having put her son through 6 years of school

SOURCE: Peterson (2000:100–101, 112–113).

---

roads have linked foresters as well as farmers to urban markets. Loggers, charcoal producers, and fuelwood collectors have mined peri-urban forests for wood, gradually creating cutover zones around cities. By the mid-1980s, the cutover zone around Lubumbashi in southeast Zaire extended 30 km from the city center, and people supplying the urban market for wood were exploiting forests over 70 km from the city center (Malaisse and Binzangi, 1985).[3]

The growing urban demand for foodstuffs sometimes leads to agricultural expansion in places far from cities. Agricultural expansion in the Genema and Lisala-Bumba regions in northeastern DRC illustrates this pattern. The drier climate on the northern fringes of the Congolian forest controls agricultural pests better than the wetter climate farther south, and rivers provide a means of transport to the urban markets of Brazzaville and Kinshasa, so small producers in the north have expanded their agricultural enterprises at the expense of the forests in order to supply distant urban markets with produce (Achard et al., 1998:48–49). Mar-

Table 6.1
Tropical Deforestation in Central Africa:
A Qualitative Comparative Analysis

A. 1980s (9/9)
    urbmkt colon fuelwood SMALLAG ROADS POP (2) +
    URBMKT colon FUELWOOD SMALLAG ROADS POP (1) +
    URBMKT colon FUELWOOD smallag roads POP (1) +
    URBMKT colon fuelwood SMALLAG ROADS pop (2) +
    URBMKT COLON fuelwood SMALLAG roads pop (1)

B. 1990s (6/6)
    logging URBMKT SMALLAG ROADS (2) +
    URBMKT SMALLAG ROADS pop (2)

The "+" in this table indicates that deforestation occurs whenever any of the listed combinations of causal conditions exists.

Upper case indicates that the causal factor is present, and lower case indicates its absence. The ratio in parentheses indicates the proportion of studies for that period whose findings agree with the Boolean expressions in the lines below it. The number in parentheses after each causal combination indicates the number of studies whose findings fit that causal conjuncture.

Factors: COLON = programs of directed settlement; FUELWOOD = wood collection for household use; LOGGING = timber exploitation, usually for export; POP = rural population increase; ROADS = construction of roads into rain forest regions; SMALLAG = smallholder agriculture; URBMKT = agricultural expansion around cities.

kets for the purchase and sale of individual plots of land have emerged for the first time in areas experiencing agricultural expansion (see box). The salience of peri-urban deforestation in the case studies analyzed in table 6.1 testifies to the importance of growing urban markets in precipitating land use transformations across the region.

Urbanization usually depletes forests in small zones around cities, but it eliminates game over a much wider area. Roads constructed by loggers open up new areas of forest for hunters who supply urban markets with bushmeat, and they often extinguish local populations of monkeys, porcupines, forest antelope, and buffalo within several years. The increased pressure from commercial hunters has altered the routines of rural residents who hunt for subsistence. After the logging roads in Gabon opened up a region for outside hunters, villagers had to travel farther into the forest from their villages before they encountered game, and the threat of extinction increased (Wilkie, Sidle, and Boundzanga, 1992; Gami and Nasi, 2001:225–227). In Gabon and the Congo, which contain approximately 80 percent of the world's remaining gorillas *(Gorilla gorilla)* and chimpanzees *(Pan troglodytes),* hunters have reduced the gorilla and chimpanzee populations by more than half since 1983 (Walsh et al., 2003).

Interior Reforestation

The patterns of deforestation summarized in table 6.1 reflect the changing trajectory of forest cover change in the region. Directed agricultural projects played a small, vestigial role in forest cover change during the 1980s (see part A of table 6.1), but by the 1990s, with the continued decline in the power of central governments, they had disappeared. During periods of political tranquility and economic prosperity, rural peoples have resettled along rivers and roads, creating corridors of cleared land in the rain forest (Wilkie and Finn, 1988). The lines of disturbed forest visible in figure 6.1 follow rivers and roads. They indicate the more intensive shifting cultivation and incomplete forest recovery that typifies road and riverside landscapes. Locations along a transportation corridor increase farmers' proximity to nonfarm labor markets. Roadside locations also enable small farmers to develop personal networks that include at least some government employees. When governments initiate agricultural development projects, smallholders mobilize these contacts to reap benefits from the new programs (Berry, 1993).

Not all new roads lead to corridors of cleared land. Loggers built networks of rough roads throughout the western reaches of the central African rain forest during the 1990s as the scale of their enterprises increased, but the construction of logging roads did not lead to the conversion of forests into farms in Congo-Brazzaville and Gabon (Wilkie, 1996; Wunder, 2003). The long distances to urban markets and the availability of relatively well paying jobs in cities reduced the number of small farmers who wanted to create farms along the recently constructed logging roads. In these instances, secondary growth enveloped the new roads and closed them to traffic within 5 years (Achard et al., 1998). In other places, logging played a familiar role in the destruction of forests. In western Zaire close to Kinshasa, the construction of logging roads initiated the familiar "follow-on" farmer sequence (Witte, 1992). Smallholders intent on producing foodstuffs for markets in ever more populous cities such as Kinshasa cleared land along roads constructed by loggers.

Logging in the region also did less to degrade the forests than it has in other parts of the tropical world. Here, as elsewhere, loggers take only selected species, but the degree of selectivity in central Africa is striking. Whereas loggers in Malaysia may harvest eight to ten species of trees from the forests and loggers close to the sea in Gabon harvest multiple species, loggers in interior Gabon only remove one species,

Figure 6.2   Two pygmies standing beside a freshly felled tree in a primary rain forest in the Democratic Republic of the Congo. Photo Credit: FAO photo (1995).

okoume *(Aucoumea klaineana)*. Okoume is a particularly light hardwood that mill owners like to use in manufacturing plywood. Because only one to two okoume occur in a hectare of forest, their removal does not disturb the forest in a massive way. The scattered nature of okoume stands also compels timber companies to exploit large areas of forest. To find the small fraction of valuable trees in the forests and mark them for cutting, logging companies employ pygmies, such as those pictured in figure 6.2, who are familiar with the forests inside concessions.

Rural people still want roads, but their ability to get them has declined. When the oil companies in Gabon began to exploit new oil fields in Gabon's coastal zone and constructed roads to serve the wells during the 1980s, village leaders requested that the companies build the new roads through their communities to gain access to urban areas (Wunder, 2003:92–93). The sparse populations of people remaining in remote areas usually do not have the numbers or organizational capacity to mount effective lobbying efforts on behalf of new roads for their regions. In the absence of a strong political appeal, government officials see little reason to build farm-to-market roads.

Appreciation in the national currency made foreign foodstuffs less expensive for urban Gabonese and encouraged them to develop a taste for inexpensive, foreign foodstuffs. Under these circumstances, trends in the markets did not induce agricultural expansion, and the government felt little need to build farm-to-market roads and expand the extent of agricultural lands to feed the Gabonese population. Without roads, affluent people in Gabon have taken to the air. In 1998, the country had 52 airports to serve a population of approximately 1 million people (Wunder, 2003:110)!

A different set of political and economic circumstances produced a similar outcome in Zaire. Mobutu's long rule by fiat made it less necessary for him to curry favor with scattered rural populations by building roads, so he neglected to do so. In addition, Mobutu allowed many existing roads to fall into disrepair. They became seasonably impassible; when roads were open, wretched conditions slowed travelers. Trips that took 1 day during the 1960s took 5 to 10 days in 1987. The political turmoil made transport more difficult. Armed groups established multiple checkpoints along roads and extracted "tolls" from people in vehicles (Draulans and Van Krunkelsven, 2002). In the midst of the political chaos, indigenous groups such as the Lese in the Ituri Forest destroyed rural roads to protect themselves. Without roads, the armed groups who controlled towns found it more difficult to exert control over the Lese in their now less accessible forest locations (Grinker, 1994:21). The political turmoil, nationalization policies, and the growing difficulty of getting agricultural products to market caused the predominantly white coffee planters in Zaire to abandon their roadside plantations during the 1970s and 1980s (Grinker, 1994:50–53). The political unrest and urban biases in expenditures also caused rural services to disintegrate after 1980. Health clinics closed, and roads built during the colonial era became impassable. Forests reoccupied old fields, but, in a pattern repeated throughout the region, the expanding forests contained few animals.

Trajectories of Change in the Central African Landscape

If reforestation in remote rural regions and deforestation in peri-urban regions together characterize forest-cover change in central Africa, then obvious questions about the net effects of these two opposed trends in land cover arise. These questions are difficult to answer. Because rural residents practice shifting cultivation in places with abundant supplies of land, they practice a much more extensive form of agriculture than the permanent agriculture practiced by smallholders in peri-urban zones where land is relatively scarce. The shifting cultivation in the interior produces relatively extensive areas of disturbed secondary forests compared with the more compact area of permanently cleared lands associated with peri-urban agriculture. The extent of the degraded forest that begins to recover in the interior exceeds the degraded forest that undergoes permanent deforestation in peri-urban zones.

This conclusion does not imply net increases in forest cover for the region, because it does not include the effects of natural increase on changes in the extent of land under cultivation. Although some populations in the region have experienced high levels of sterility (Bailey and Aunger, 1995) and a growing incidence of infection with the human immunodeficiency virus (HIV), the region's population grew by an average of 3.0 percent per annum, and the urban population more than doubled between 1980 and 1999 (World Bank, 2001).[4] Smallholders met at least a portion of the corresponding increase in demand for foodstuffs and fuelwood by farming more land and extracting more wood from the remaining forests.

The locations of the new farm fields shifted with changes in economic conditions. During the periods of economic prosperity that attend oil booms or during periods of political turmoil in the countryside, people congregate in urban areas, and peri-urban agriculture expands around the cities. During periods of economic stagnation, as occurred in Cameroon after the sharp decline in oil prices in 1986, some recent migrants to the cities return to rural areas, and the practice of subsistence agriculture expands in remote rural areas (Mertens and Lambin, 2000). In this last circumstance, natural increase in rural households may determine the pace at which shifting cultivators expand into the forests (Wunder, 2003:209).

Although the physical barriers to the exploitation of the central African forests will not disappear in the near future, the regional pattern of peri-urban deforestation and interior reforestation should change in the coming decades as international corporations exhaust the easily exploited deposits of oil, diamonds, and copper. If political unrest subsides enough to permit resettlement and investment in the countryside, the reduced receipts from mining should magnify the importance of logging and farming in the region. To the extent that these rural enterprises become economically important, people may find more economic opportunities in rural areas. The accompanying increase in human pressures on the rain forests may be countered to some extent by the continuing partition of the region into zones dominated by loggers, tribal groups, and park guards.

Foreign logging firms, which hold most of the concessions to exploit sparsely populated sectors of the forest in the west, have engaged in purely extractive activities, harvesting economically valuable old-growth trees with no concern for how regeneration might occur (FAO, 2003b). As these firms deplete old-growth species, they will no doubt show more interest in harvesting a wider range of species from their concessions, but they may also show some interest in replenishment planting, especially if their activities make it possible to sell their timber as certified wood. Several features of this situation suggest that policies designed to promote sustainable logging through certification schemes may help limit the impact of logging on the forests of the region. Unlike many small, locally owned logging firms, the multinational firms can afford the costs of certification, and in European markets they obtain a competitive advantage if they log in a certifiably sustainable way (Atyi and Simula, 2002). The development of a national certification scheme in at least one country in the region (Cameroon) indicates some local interest in certifying and therefore controlling the damage that logging operations do to the forests. This direction in forest management assumes that concession holders can exercise meaningful control over timber harvesting within their concessions.

Land reforms that provide forest people such as the Efe pygmies and Lese horticulturalists with tenure over the forested lands that sustain them also offer some conservation benefits. If shifting cultivators, foragers, and hunters acquire ownership over areas such as the Ituri Forest in the northeastern region of the DRC, they could abolish the open access that currently encourages unsustainable exploitation (Bailey, 1996). This policy would give local residents an incentive to preserve local game populations in the face of hunting pressures from urban-based hunters, and it would provide local populations with a stream of income from the sale of timber.

"So many parks and so little funding for them" (Wilkie, Carpenter, and Zhang, 2001:691)! This phrase sums up the dilemma faced by conservation groups dedicated to preserving biodiversity in central Africa. The insufficient levels of funding call for prioritizing the parks most deserving of foreign-financed protection and reconfiguring the other protected areas as multiple-use reserves under local control. This policy direction assumes that local stakeholders can establish effective arrangements for governing the use of natural resources through some sort of adaptive co-management.

None of these policies will do much to alleviate the overwhelming poverty experienced by the regions' inhabitants. Debt relief for national governments would contribute to this end by providing governments with the financial resources to deliver vital services such as education to their citizens. Given the urban biases in government expenditures, this type of policy would also discourage return migration to rural areas. A revision in the Clean Development Mechanisms in the Kyoto Protocol to permit carbon sequestration payments for carbon that accumulates in old-growth forests would provide ecological and economic benefits for large numbers of impoverished inhabitants of rain forest regions. These initiatives, such as the locally based ones outlined here, assume the emergence of effective institutions for managing natural resources. The multiple paths to biodiversity conservation in central Africa will go nowhere without more effective political institutions in local and national governments.

7

East Africa: Sustainable Spots Surrounded by Degrading Expanses

We proceed westwards, and a little south through a country covered with forest trees, thickly planted, but small.... A good portion of the trees of the country have been cut down for charcoal, and those that now spring up are small; certain fruit trees alone are left. The long slopes of the undulating country, clothed with fresh foliage, look very beautiful. The young trees alternate with patches of yellow grass not yet burned; the hills are covered with a thick mantle of small green trees with, as usual, large ones at intervals.... The Mopane forest is perfectly level, and after it rains the water stands in pools; but during most of the year it is dry. The trees here were very large, and planted some twenty and thirty yards apart; as there are no branches on their lower parts, animals see very far.
DAVID LIVINGSTONE, *The Last Journals of David Livingstone* (1970 [1874]:142, 147, 160)

The dry forest described by Livingston covers most of East Africa.[1] Called miombo woodlands, it stretches for 1,500 hundred miles, east and north from eastern Angola across Zimbabwe, Zambia, the southeastern Democratic Republic of the Congo (DRC), and Malawi, to northern Tanzania.

Figure 7.1   Slash and burn in a miombo woodland, Zambia. Photo credit: P. Lowery (1996), FAO photo.

It is the largest deciduous, dry forest in the world. It contains less biomass than the closed forests of the humid tropics. Two genera of trees, *Brachystegia* and *Julbernardia,* dominate the miombo woodlands. They rise to between 12 and 15 m in height, sprout green leaves during the wet season, and shed them during the dry season. The trees are usually spaced regularly, but sometimes they occur in thickets. In most places, they do not achieve a closed canopy. Contemporary miombo woodlands exhibit the marks of human intervention, most frequently in the form of heavily coppiced tree trunks (figure 7.1). A second type of dry forest, dominated by *Acacia* species, occurs to the north of miombo woodlands at lower elevations in Kenya, northern Tanzania, the Sudan, and Ethiopia. Trees in these dry forests and savanna woodlands frequently do not attain the 12- to 15-m height of the miombo woodlands and contain less biomass than trees in miombo woodlands (Allen, 1985; Dewees, 1994).

Other types of forests exist in the microclimates created by the volcanoes that emerge from the East African plains. The humid "sky islands" that ring the volcanoes support moist forests in the Eastern Arc mountains in Tanzania and Kenya.[2] Montane forests also surround the volcanic cones that extend north along the Albertine riff in the eastern

DRC, Burundi, Rwanda, and Uganda. Patches of humid closed forest still exist at middle elevations, about 2,000 m above sea level, on the plateau stretching west from Lake Victoria in Uganda and along ridges in Rwanda and Burundi. These forests represent the easternmost extension of the Guineo-Congolian forest that extends inland across central Africa from the Atlantic Ocean. Figure 7.2 maps the forests in the region.

East Africans, to a greater degree than central or West Africans, live in rural places. Although urban areas have grown rapidly since 1980, the region's population has remained predominately rural. In 1980, 86.2 percent of the region's population resided in rural areas; in 1999 76.3 percent of the population still resided in rural areas. Rates of population increase in the region remain high, ranging from 2.5 percent per annum in Rwanda and Burundi to 3.0 percent per annum in Kenya for the 1980 to 1999 period (World Bank, 2001). Whereas central Africa with its large forests and small rural populations contains large amounts of woody biomass per capita, East Africa with its sparse or small forests and large rural populations contains little woody biomass per capita.

East Africans had the lowest per capita incomes in 1999 among the people of tropical Africa. Their incomes averaged $250 compared with $281 for central Africans, and $330 for West Africans (World Bank, 2001). The economic fortunes of East Africans deteriorated during the last two decades of the twentieth century. People lost jobs in the mines when prices for ore declined and company executives decided not to re-invest in the mines. Structural adjustment policies imposed on indebted governments by the International Monetary Fund prevented new initiatives in the public sector in Zambia, Malawi, and Tanzania. Rural roads deteriorated when, in the midst of fiscal crises, government officials cut the budgets for road maintenance. Political turmoil and violence stifled economies in Rwanda, Burundi, and Uganda; similar problems afflicted the economies of Kenya and Zimbabwe to a lesser degree.

Although large landholdings have increased in recent years in Kenya (Berry, 1993:19–20) and remain a salient, though contested, part of agriculture in Zimbabwe, most producers in the region work small plots of land on either a permanent or a temporary basis. People maintain individual plots of land in peri-urban districts and along roads in remote rural regions (Moore and Vaughan, 1994:35–36). Away from the roads, farmers engage in shifting cultivation on communal lands. In the citemene system of shifting cultivation used by Bemba cultivators in northern Zambia, farmers coppice trees in the miombo woodlands,

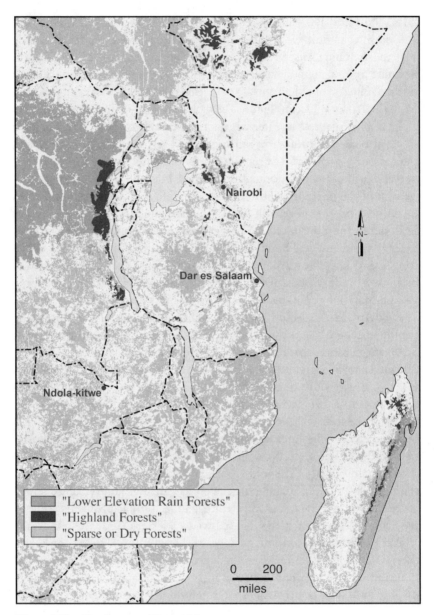

Figure 7.2   The forests of East Africa.

stack the cuttings in a small area, burn the accumulated wood, and then cultivate the ash-enriched soils for two to three seasons. Some small-holders, such as tobacco farmers in Malawi, produce for export, but the overwhelming majority of smallholders either produce for local urban markets or for household consumption. Poor rural residents forage in the woods around their villages for a large proportion of their liveli-hoods. Farm households in the miombo woodlands of Tanzania earned 58 percent of their household incomes from the sale of gathered products such as honey, wild fruits, and wood processed into charcoal (Monela et al., 1999). In Zimbabwe, poor households in rural areas earn 40 percent of their household income from forest products such as thatching grass, wood, and game (Cavendish, 2000).

Against this somber political-economic backdrop, the regional rate of deforestation climbed slightly during the last two decades of the twen-tieth century, from 0.9 percent per annum in the 1980s to 1.1 percent in the 1990s (FAO, 1993, 2001). This small change in aggregate rates conceals sharp intraregional variations in forest cover trends. Although forest cover may have actually increased in densely populated rural areas with good transportation infrastructure, it declined or degraded in more peripheral places. Old-growth forests also declined in extent whenever political turmoil broke out in a place. The following pages outline an explanation for these spatial and temporal patterns in the forests of East Africa, beginning with the changes introduced in the early twentieth cen-tury by colonial regimes.

Forests in the Precolonial, Colonial, and Immediate Postcolonial Eras

Successive waves of exploitation and abandonment characterized the precolonial history of the region's forests. During the eighteenth and nineteenth centuries, traders from coastal settlements such as Zanzibar sought ivory and slaves in the interior. Inland trading centers sprang up to facilitate the commerce, and the inhabitants exploited the nearby for-ests for fuelwood and timber. In the southern miombo woodlands, indig-enous peoples cleared tracts of forested land around fortified settlements to defend themselves better against Ngoni invasions. The forests recov-ered after a rinderpest epidemic in the late nineteenth century killed 90 percent of the cattle in the region (Misana, Mung'ong'o, and Mukamuri,

1996:74–80). The forests found by the first colonial administrators bore the marks of these earlier human activities.

The imposition of colonial rule in the late nineteenth century triggered large changes in the East African landscape. In the highlands of Kenya and Tanganyika, European settlers seized native lands and set up plantations while colonial officials, avowing an interest in conservation, declared forests and dryland areas off-limits to native peoples. These coercive measures squeezed growing indigenous populations into small reserves, which they and their herds soon stripped of almost all vegetation (Tiffen, Mortimore, and Gichuki, 1994:3–5; Barraclough and Ghimire, 1996). Railroad construction from the coast into the interior expanded the areas supplying urban markets with wood and spurred deforestation in corridors along the rail lines in Tanganyika, Kenya, and Uganda (Castro, 1988; Barraclough and Ghimire, 1996).

In the more arid, lower-elevation regions, colonial officials initiated the first of a series of efforts to resettle indigenous peoples. Between 1900 and 1920, agents of the British South African Trading Company in northern Rhodesia tried to concentrate Bemba tribesmen in communities along roads to facilitate the collection of taxes (Moore and Vaughan, 1994). In an attempt to quell growing political unrest after World War II, colonial officials in Kenya established settlements for smallholders on sparsely populated lands with dry forests that they had declared off-limits to indigenous peoples early in the twentieth century (Tiffen, Mortimore, and Gichuki, 1994:72). Colonial officials in Northern Rhodesia established similar resettlement schemes in a less tumultuous political environment during the 1940s and 1950s (Moore and Vaughan, 1994:121–128).

After independence, the new governments continued this policy of planned landscape transformation. To incorporate rural Tanzanians into national politics and deliver services to them more easily, Julius Nyerere tried to concentrate them in "ujamaa" villages during the 1960s and early 1970s (Barraclough and Ghimire, 1996:106). Zambia's government promoted similar "village regroupment" schemes during the early 1970s (Moore and Vaughan, 1994:137). All of these schemes encouraged people to abandon shifting cultivation in favor of practicing permanent agriculture on fields close to villages and roads. The planners and aid officials who promoted these projects were driven by the twin specters of hunger and continued rural poverty (Allan, 1965). The settlement schemes usually proved to be unworkable and expensive, so they either had short lives, like Nyerere's "villagization" program, or they never expanded beyond "demonstration farms," such as the Mungwi scheme in Zambia.

Although colonial officials did not succeed in relocating most indigenous peoples, the officials did set in motion some far-reaching changes in the landscape when they began to promote the cultivation of cassava during the 1930s (Moore and Vaughn, 1994:35–36). After World War II, large numbers of indigenous peoples began to cultivate cassava on semi-permanent plots in Tanzania and Zambia. Villagers increased their food supply substantially by adopting cassava as a staple, but they never completely abandoned the cultivation and consumption of millet grown through the traditional "citemene" system of shifting cultivation (Moore and Vaughan, 1994). Smallholders never converted more than 1 or, at most, 2 hectares of woodlands into cassava fields, but the aggregate effect of their efforts on the extent of forests may have been considerable.[3]

Although the imposition of colonial rule and the associated economic transformations decimated the region's forest resources, an alternative pattern of landscape change—some might call it "environmental recovery"—emerged after 1930 in the most populated, highland districts in the region. Kikuyu smallholders in the highland districts surrounding Mt. Kenya refused to participate in the development of communal forest plots, but they did begin to plant commercially valuable trees on their small, private landholdings after nurseries began to make seedlings available during the 1920s. The Kikuyu accelerated their tree planting after 1930 when they found that they could sell the bark of wattle trees *(Acacia mearnsii)*, an Australian exotic, to exporters who used it to process leather. The tree planting continued at a rapid pace until the Mau Mau uprising in the early 1950s, when British troops destroyed the wattle forests on smallholdings as a security measure (Castro, 1996:128–139). Although this early example of afforestation did not survive the political turmoil surrounding independence, the stories about it provided postindependence smallholders with an example of how they might profit from agroforestry.

Trajectories of Change After 1980

Table 7.1 summarizes more recent changes in the East African landscape through a qualitative comparative analysis of 25 studies of land-cover change in East Africa between 1980 and 2000. The case studies describe and explain land-cover changes in small areas of Ethiopia, Kenya, Mada-

Table 7.1
Deforestation in East Africa: A Qualitative Comparative Analysis

A: 1980s and Earlier (12/12)
  urbmkt FUELWOOD POP SMALLAG (1) +
  high urbmkt POP SMALLAG (5) +
  high urbmkt FUELWOOD SMALLAG (2) +
  high urbmkt FUELWOOD POP (1) +
  high FUELWOOD POP smallag (2)

B: 1990s (11/13)
  high URBMKT FUELWOOD POP SMALLAG TENURE (2) +
  HIGH urbmkt fuelwood pop SMALLAG TENURE (1) +
  high fuelwood POP SMALLAG tenure (4) +
  high urbmkt FUELWOOD pop SMALLAG tenure (1)

The "+" in this table indicates that deforestation occurs whenever any of the listed combinations of causal conditions exists.

Upper case indicates that the causal factor is present; lower case indicates its absence. The ratio in parentheses indicates the proportion of studies for that period whose findings agree with the Boolean expressions in the lines below it. The number in parentheses after each causal combination indicates the number of studies whose findings fit that causal conjuncture.

Factors: FUELWOOD = extraction of wood products (e.g., fuel, poles); HIGH = highland location; POP = resource-degrading population growth; SMALLAG = smallholder agriculture; TENURE = insecurity in land tenure; URBMKT = deforestation in peri-urban areas.

gascar, Malawi, Tanzania, Uganda, and Zambia. The causal configurations in parts A and B of the table indicate sharp geographic differences, but few historical changes, in the forces driving deforestation during the two decades. FUELWOOD figures prominently in the causal configurations, because the region contains large numbers of poor, rural people who derive significant amounts of their income from forest products that they collect in the region's sparse forests and open woodlands. The salience of the SMALLAG and POP terms reflects the importance of agricultural expansion by growing populations of smallholders in processes of forest cover decline. Political turmoil also contributed to deforestation by undermining land tenure (TENURE) systems during the 1980s in Uganda and during the 1990s in Rwanda and Burundi.[4] The resulting "anything goes," open-access situations encouraged rapid exploitation of the standing forests with little thought for future yields or global concerns about biodiversity.

Both the 1980s and the 1990s analyses indicate sharp intraregional differences in land-cover change. Some farmers converted lowland dry forests into fields, and others planted trees in the highland regions of Kenya and Tanzania. People continued to exploit forests away from the roads during the last two decades of the century, but they had cut almost all of the harvestable wood in peri-urban places and places along all-weather roads earlier in the century. In response to scarcities of wood in more accessible areas, farmers in these places have begun to plant fruit- and timber-bearing trees on their farms. The following sections describe these different processes of landscape change in more detail.

Markets, States, and Environmental Stabilization in Densely Populated Places

During the last decades of the twentieth century, the women who worked the densely settled agricultural landscape of the East Africa's humid up-lands planted more trees than they had in previous decades. Their work gave the landscape an almost forested appearance in places where small farms predominated. In Kenya's Machakos district, small landholdings contained an average of 59 fruit trees per hectare compared with 21 fruit trees on large landholdings. Machakos smallholders also planted fuel- and timber-producing trees on their farms, most frequently *Eucalyptus* species and *Croton megalocarpus*. Other environmental improvements accompanied the increase in tree planting in Machakos. To reduce soil erosion on sloped lands where they cultivated vegetables, farmers created terraces. To conduct water from roads to fields and to prevent soil deposits on fields beneath ridges, farmers dug trenches along the edges of their fields and lined them with banana plants and fruit trees (Tiffen, Mortimore, and Gichuki, 1994:178–225).

Growth in the size and increase in the wealth of nearby urban populations explains the proliferation of trees and other environmental improvements on Machakos farms. The borders of the district begin immediately to the east of Kenya's largest city, Nairobi (2000 population, 1.8 million), and the main road between Nairobi and Kenya's second city, Mombasa, runs through the district. Beginning in the 1950s, Machakos smallholders found that they could profitably grow fruit, vegetables, and coffee for urban markets in Nairobi, Mombasa, and overseas. The growth of the cities also created opportunities in the nonfarm sector for Machakos residents, both as day laborers in district towns and as temporary labor migrants to the cities. In many instances, household heads used the earnings from the nonfarm jobs to finance environmental improvements on the farms.

The relative political tranquility that prevailed in Kenya between 1965 and the early 1990s enabled this process of agricultural development by

---

Planting Trees on Ali's Family Farm: Afforestation and Conflicts over Land

Ali's family lives in the Shimba Hills of the coastal plain in southern Kenya. Beginning in the 1920s, the Mijikenda peoples of this region abandoned cattle herding, began planting coconut trees, and settled in homes built in the small groves of coconut and, later, cashew trees that they had planted. Through their planting, the Mijikenda changed the regional landscape over three or four decades from scrub growth to a dry, fruit-bearing forest. The members of Ali's family are Muslims from the Digo subgroup of the Mijikenda, and they, like others, established a smallholding in the region during the middle of the twentieth century. When Ali's grandfather died, Ali's father inherited some of the family land, and in 1983 he decided to plant trees on it. The tree planting sparked a dispute over the inheritance. Ali's grandfather had had 12 children by three wives, and each wife maintained a separate household with her children. Ali's father, as the only child of one of the wives, stood to inherit a sizable tract of land if, following the dictates of Digo customary law, each household inherited a tract of land. Ali's uncles and cousins argued for dividing the land up by child, following accepted practices in Islamic law. This arrangement would have left Ali's father with one twelfth of the grandfather's land. The dispute went to court, and conflict within the family sharpened. Eventually, the half-brothers stopped speaking to one another. In 1985, the court ruled in favor of division by household. The history of this dispute underscores the tight connection between tree planting and land tenure security.

SOURCES: Parkin (1972); Ng'weno (2001:124–125).

---

making it possible for small farmers to secure claims to the parcels of land that they cultivated. After 1965, farmers began registering their claims to land. Although only 35 percent of them had completed the registration process by 1991, initiating the process and obtaining the document that comes with it probably strengthened their claims to land and removed a potential impediment to investing in the land (Tiffen, Mortimore, and Gichuki, 1994:65–66). With deeds or related documents in hand, many smallholders began to invest in their land by planting trees. As Ali's story (see box) illustrates, planting trees can also become a strategy for strengthening claims to land.

All of the humid, highland districts in Kenya appear to be experiencing an increase in farm forests. A study that oversampled lands close to roads in the Kenyan highlands used aerial photographs from 1986 and 1992 to calculate changes in woody biomass and found a 4.7 percent annual increase in woody biomass during the 6-year period. Almost all of the increase in tree cover occurred on small farms (Holmgren, Masakha, and Sjoholm, 1994). Similar trends in tree cover characterized highland districts in Rwanda before the genocide of the mid-1990s. More than 60 percent of the respondents reported that they had increased the number of trees that they had planted during the 1980s (Kampayana, 1992). A similar story of increases in tree cover has unfolded in the Western Usambara mountains of Tanzania. After colonial officials limited access to forest reserves during the 1930s, smallholders intensified agriculture. They began producing fruits and vegetables for sale in the urban markets of Arusha and Dar es Salaam. Peasants planted some trees to produce fruit for sale, and they planted other trees to meet their households' needs for food, fuel, and construction materials (Barraclough and Ghimire, 1996:104–105).

Smallholders in accessible, densely populated areas of East Africa evidently practice what R. Mac Netting (1993) called "the ecology of small scale, sustainable agriculture." When market conditions permit, they gradually transform a denuded agricultural landscape into a wooded agricultural landscape by planting trees on farms. These changes constitute environmental stabilization rather than ecological restoration, because the biodiversity of the newly planted woodlands does not compare with the biodiversity of the plant and animal communities eliminated during the colonial era. In effect, farmers have "domesticated" the landscape.

The gradual extension of state control over rural districts during periods of political calm, combined with intervention by foreign environmentalists, may have gradually improved the degree of protection afforded to the small patches of humid, moist forests that remain in the region. Although environmentalists often decry the ineffectiveness of recently created parks in protecting biodiversity, the creation of a park may at least slow down the local rate of deforestation. The creation of the Tana River National Primate Reserve in eastern Kenya offers a case in point. During the 15 years prior to the creation of the reserve, from 1960 to 1975, the gallery forests along the Tana River declined 56 percent in extent. In the 12 years after the creation of the reserve, from 1976 to 1988, the forests declined only 4 percent in extent (Medley, 1998:48–49).

The struggle to preserve the habitat of mountain gorillas *(Gorilla gorilla beringei)* on the slopes of the Virunga volcanoes in Rwanda demonstrated that the preservation of forests within parks often depends on the actions of a committed park staff. Belgian colonial administrators created the Albert National Park around the volcanoes in 1929. Political and economic pressures to accommodate growing populations of smallholders spawned agricultural development projects that reduced the size of the reserve during the next 50 years. Colonial authorities took 18,500 hectares for an agricultural settlement scheme in 1958. Ten years later, the government of Rwanda appropriated another 25,000 hectares for an agricultural development project. During the late 1970s, planners proposed to take another 12,000 hectares of land for agriculture. The 1970s plan would have split the reserve into small islands of forests on the lower slopes of each volcano, thereby isolating the already small populations of gorillas from one another. To counter this plan and other efforts to destroy the forests, a coalition of foreign primatologists and reserve employees created the Mountain Gorilla Project. During the 1980s, the project's staff demonstrated that visits to mountain gorillas by foreign tourists could generate a considerable amount of foreign exchange for Rwanda (Adams and McShane, 1992:184–199). The economic promise of the tourist trade and courageous work by the project's Rwandan staff prevented further losses of land and wildlife during the political turmoil and genocide of the mid-1990s (Vedder et al., 2001:561).

In the Kibale forest in Uganda, just to the east of the Virunga, logging financed by Asian merchants and poaching by local residents decimated local wildlife populations during the 1960s. The political chaos experienced during the Amin–Obote regimes in the 1970s and 1980s reduced the logging but increased the poaching inside the reserve. The restoration of political stability in the early 1990s strengthened conservation efforts in the reserve. The government evicted several thousand squatters from the southern portions of the reserve in 1992, and poaching declined. A local infrastructure of individuals and organizations dedicated to conserving the forest and its wildlife also emerged with the restoration of political stability in Uganda (Struhsaker, 1997:9–14). The presence or absence of local coalitions of conservation-minded people played a crucial role in determining whether or not five other Ugandan reserves remained relatively undisturbed or experienced invasion and disturbance during the 1980s and 1990s (Banana and Gombya-Ssembajjwe, 2000). In the protected reserves, as in the highland agricultural zones, environmental stabilization has occurred.

Resource Depletion and Forest Decline in Peripheral Places

In peripheral parts of East Africa, farther from urban places but still accessible to their markets, forests continue to decline in extent. Unlike the long-settled, sparsely forested highlands where smallholders have begun planting fruit trees, these more peripheral, lower-elevation lands often contain fairly extensive dry forests. Until recently, many of these forests have enjoyed the passive protections afforded by inaccessibility (Arnold et al., 2003:22). The recent expansion of urban markets and the construction of farm-to-market roads in some places has created new opportunities for trade in wood products, which has increased the volume of cutting in dry forests. Fisherfolk on the shores of Lake Malawi cut the dry forests to feed the furnaces in which they smoke fish for sale in urban markets (Abbot and Homewood, 1999). In Tanzania in the delta of the Rufiji River, local people have begun to extract large amounts of wood from mangrove and other forests to sell in the markets of Dar es Salaam, 150 miles to the north (Barraclough and Ghimire, 1996:105, 107). In these places, people acquire wealth by decimating local supplies of natural resources. They rarely have secure rights to these resources, so they hasten to harvest them. At least some of these people may then be able to use their natural resource–derived wealth to begin nonfarm enterprises that offer them a way out of rural poverty (Barbier, 1993).

Resource-Degrading Poverty Traps in Peripheral Places

People in other places, oftentimes even farther from urban areas, cannot follow the resource depletion path out of poverty. Instead they find themselves caught up in resource-degrading poverty traps. Land-use changes in a rural district of the Ethiopian highlands between the late 1950s and early 1990s illustrate the pattern. To accommodate increases in the district's population, farmers extended the area under cultivation and households collected more firewood from the surrounding forests and shrublands. Farmers could not afford fertilizers for their lands, and they frequently used animal dung for fuel, so they had nothing with which to fertilize their lands. Under these circumstances, they used each piece of land for less time before fallowing it, and they allowed lands to lie fallow for shorter periods of time before reusing them. With these practices, shrublands and forest soon degraded to barren ground. Local labor markets offered little or no off-farm employment, so smallholders could not supplement their income from agriculture or use income from the non-farm sector to invest in productivity-enhancing conservation activities such as tree planting. Under these circumstances, forests and shrublands declined between 30 percent and 50 percent in extent, whereas open, barren ground and urban settlements expanded by 332 percent and 194 percent, respectively (Tekle and Hedlund, 2000).

Trends in the larger economy can exacerbate this grim situation. Declines in government receipts from the mining of copper in Zambia during the 1970s and 1980s precipitated a debt crisis for the national government. As part of an austerity plan authored by the International Monetary Fund, the government agreed to end its subsidies for fertilizers and the transport of agricultural commodities. The cutbacks coincided with massive layoffs at the copper mines. After receiving their layoff notices, some workers and their families decided to return to their rural communities of origin. To provide for themselves, the returning migrants and many of their rural neighbors resumed the citemene system of shifting cultivation on marginal lands during the late 1980s and early 1990s. In the now more populous rural districts of northeastern Zambia, the growing reliance on citemene increased the rates of deforestation and degradation in the region's dry forests (Misana, Mung'ong'o, and Mukamuri, 1996:88; Holden, 1997).

Tenure insecurity in the same places often accelerates the loss of forests. Weak states find it difficult to provide state services for registering claims to land, especially in remote rural districts.[5] When wars, civil wars, or spasms of collective violence test the capacities of weak states, most forests lose their public protection, and deforestation rates accelerate. Table 7.2 documents a sharp increase in deforestation rates in Burundi and Rwanda from the 1980s to the 1990s, an increase almost certainly attributable, either directly or indirectly, to the intensified political struggles between ethnic groups in both countries.

Three associated conditions define resource-degradation poverty traps in which people have little choice but to destroy the forests to sustain themselves (McPeak and Barrett, 2001). First, smallholders have no capital to finance more intensive agriculture. People have little or no access to nonfarm labor markets, so no one has savings from nonfarm jobs to invest in agriculture. Second, people have poor access to markets, so

Table 7.2
Political Turmoil and Forest-Cover Change:
Burundi and Rwanda During the 1980s and 1990s

	1980–1990 (% Change per Year)	1990–2000 (% Change per Year)
Burundi	–0.6	–9.0
Rwanda	–0.3	–3.9

SOURCES: Food and Agricultural Organization (FAO). 1993. "Forest Resources Assessment, 1990—Tropical Countries." FAO Forestry Paper, #112. Rome; and FAO. "Forest Resources Assessment, 2000." FAO Forestry Paper #140. Rome.

they have little incentive to intensify production. Third, a fluctuating, uncertain institutional context makes land tenure arrangements unstable, so the poor have few incentives to think in terms of long-term investments in the lands and forests around them. Unlike the fisherfolk who cut wood to smoke fish for urban markets, or the peasants who cut down trees for sale in urban markets, these people have little hope of accumulating wealth through the exploitation of natural resources, so they do not have a pathway out of poverty. Instead, they meet their subsistence needs through the unsustainable extraction of resources from deteriorating lands and forests. In this sense, they find themselves trapped.

Conclusion: Contrasting Trajectories of Land-Cover Change

Although environmental recovery may characterize more accessible, peri-urban places, the widespread, debt-induced cutbacks in services to small farmers, coupled with the persistence of citemene in Zambia and the continued settlement of dry forests in lowland Kenya (Tiffen, Mortimore, and Gichuki, 1994) and Rwanda (Kampayana, 1992), suggest that many rural households in remote places are caught up in resource-degrading poverty traps. Figure 7.3 outlines the spatial organization of these contrasting trajectories of landscape change. The changes in landscapes undoubtedly varied from year to year with fluctuations in political stability and economic growth in the region.

The contrasting trajectories of landscape change within the region suggest a dual strategy for policymakers: debt relief for governments and community natural resource management for villagers. These policy initiatives should have distinct spatial effects. Debt relief could assist small-

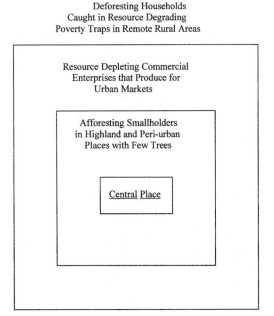

Figure 7.3   The spatial order of livelihoods and landscape change in East Africa.

holders in densely populated and peri-urban areas through the provision of earmarked funds for credit and extension services to aid the already high level of farm forestry in East Africa. By facilitating farm forestry, this policy change would promote the geographic expansion of the environmentally stabilizing trends apparent in Machakos and the humid highlands. Because small tree plantations require relatively little labor, even households that have someone working in an urban or mining center may be able to participate in these programs. The passage of forest reforms in many East African countries during the 1990s should reinforce the investment-inducing effects of easier credit by strengthening smallholder and community claims to local forest resources. These changes, coupled with the scarcity of forest resources in the region, have provided the impetus behind the recent spread of farm forestry in the region.

Community-based natural resource management (CBNRM) should primarily benefit remote rural areas. Because East African rural populations in remote areas rely heavily on common (pool) resources such as game and other forest products for their sustenance (Wunder, 2001),

CBNRM would appear to offer a potential way to prevent further declines in these resources. State-assisted CBNRM would curtail the open-access conditions that prevail in so many remote rural areas and allow the poor to fashion sustainable patterns of foraging for forest products. The presence of local afforestation programs administered by community elders appears to explain why some remote rural communities in Ethiopia experienced net increases in forests, whereas other, equally remote communities without programs lost forests (Bewket, 2002; Tekle and Hedlund, 2000). As the brief history of the CAMPFIRE program in Zimbabwe shows (Child, 2000), the creation of these programs does not ensure that communities will acquire the power to shape conservation programs or enjoy the real economic returns to conservation. Problems of implementation, not substance, have hampered this program. Big game represents the most lucrative of common (pool) resources in the region, so more centralized institutions such as corporations, government agencies, and tribal councils had reason to wrest control of CAMPFIRE hunting revenues from the villages. These political struggles do not, however, invalidate the CBNRM idea; instead, they underscore the importance of village-level political mobilization to ensure the implementation of these plans.

Even with these policies successfully implemented, the regions' biodiversity will remain at risk. Rural residents may decide against forest conservation, and farm agroforestry in the highlands will do little to restore the biodiversity of the forests that occupied these sites before the colonial era. Still, the environmental stabilization promised by these policy directions remains a worthwhile goal when denuded landscapes or degraded forests are the alternative.

8

South Asia: A Turning Point for Forests?

The situation of the village is very beautiful. The low hills around, some partially cleared for cultivation, some entirely so, ... appear in fine contrast with the dark tints of the lofty mountains of Moyur and Yaloo, and others more distant that surround it.... The road from this to Tema's village, which is about two miles distant and northwest of this village, continues over low hills, many of which have been cleared and are now fallow, and after a time will be taken up. Between the villages barricades are constructed in different places to keep the Myttons (horned cattle) from the cultivation when necessary.
E. Dalton, *An Excursion Up the Subansiri* (1959 [1845]:145–146)

Unlike other tropical regions, south Asia has long had densely settled rural areas such as this one described by a nineteenth-century British traveler. People sustained themselves first through shifting cultivation and then through permanent cultivation in the more densely populated areas. Over time, they destroyed most of the forests in lowland regions. Long before the British established colonial rule in India, villages had established local traditions of self-governance (Richards, 1987:301–302; Guha, 2000:21). Village leaders revived these traditions during the last two decades of the twentieth century when, with encouragement from federal forestry officials, they reestablished local control over some small forests. In this way, south Asian villages became "the cradle of modern community forestry" (Stone and D'Andrea, 2001:8). This shift in the politics of forest governance has begun to produce changes in both forest cover and forest composition in south Asia.

Four countries—Bangladesh, India, Nepal, and Sri Lanka—constitute the region of south Asia. Figure 8.1 maps the forests of the region. Dry, monsoon-influenced, sal *(Shorea robusta)*-dominated forests cover much of the interior in peninsular India and Sri Lanka. Along the western edges of the continent and extending south into Sri Lanka, small islands of moist forest persist on the ridges and upper flanks of the Ghats mountains. Like the forests of Southeast Asia, the moist forests of western India contain trees of great commercial value, including teak *(Tectona grandis)* and Indian rosewood *(Dalbergia latifolia)*. Loggers eliminated almost all stands of these trees during the colonial period. The northern edges of the subcontinent also contain forests on the south-facing slopes of the Himalaya and at the base of the Himalaya in the Terai district of Nepal.

As the map in figure 8.1 suggests, the tropical rain forests of south Asia are smaller and much more fragmented than the forests of South America or central Africa. For these reasons, a much larger proportion of south Asian forests exist close to fields, pastures, and villages. These spots of intense human activity spill over into forests and create edge effects that degrade forests. Fires started by shifting cultivators spread into the adjacent woods. Scrub growth and fields between patches of forest prevent seed dispersal and discourage animal traffic, which in turn makes it more difficult to maintain biodiversity in each forest fragment. People from villages lop branches off trees, gather litter from the forest floor, and cut down an occasional tree to provide both fuel and fodder for themselves and their livestock. Over time, these activities thin the canopy of a forest, converting it from a dense to an open woodland with much less biomass (Sen et al., 2002). Although forests covered 19.5 percent of India's land area in 1994, approximately 40 percent of these forests had thin crown covers of between 10 and 40 percent (Ravindranath and Hall, 1994:521). Only 2 percent of the country's land area contained primary forest in the early 1980s (Guha, Prasad, and Gadgil, 1984:246). Under these circumstances, trends in forest cover may not be the most revealing indicator of trends in biodiversity. Forest cover may

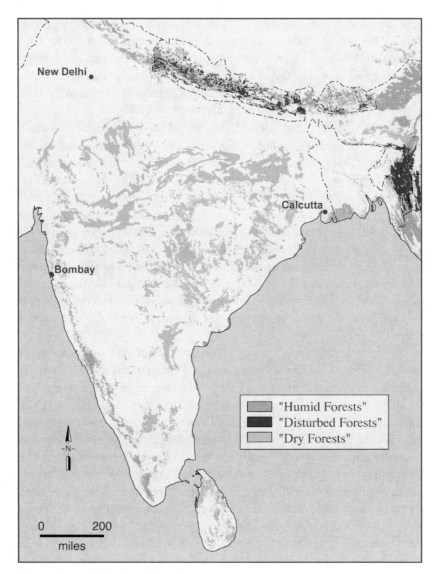

Figure 8.1   The forests of south Asia.

remain stable while the forests undergo a slow process of degradation. Although almost all of the forests in the region suffer from some degradation, it is, as indicated in figure 8.1, particularly acute in the foothills of the Himalaya (Collins, Sayer, and Whitmore, 1991:127).

In south Asia, large numbers of people use products from small forests. In 1990, for every 1,000 persons, Southeast Asia contained 8.1 hectares of forest, Latin America, 20.6 hectares of forest, and Africa, 24.8 hectares of forest. In contrast, south Asia contained only 0.5 hectares of forest for every 1,000 persons (Ravindranath and Hall, 1994). Although the proportion of the region's population living in rural areas declined from 78.4 to 72.9 percent between 1980 and 1999, large numbers of south Asians continued to live in close proximity to forests. The regional population continued to grow, albeit at slower rates than during the 1960s and 1970s. At the turn of the century, population growth rates per year ranged from 1.3 percent in India to 2.5 percent in Nepal. The vast majority of the region's residents experience lives of grinding poverty. Per capita annual income averaged $434 in 1999 (World Bank, 2001). Under these circumstances, forest products offered many people an important source of both income and subsistence, so the human pressures on many south Asian forests have been intense for the past half century.

Against this tableau of a poor, predominantly rural population for whom forests provide vital daily resources, the changes in forest cover during the 1990s, as reported by the Food and Agricultural Organization (FAO) in table 8.1, seem somewhat surprising. Both Bangladesh and India experienced "turnarounds" in forest cover trends, with either net afforestation or little change in forest cover occurring for the first time in decades. The continued decline in forest cover in Nepal and Sri Lanka suggest the persistence of earlier patterns of deforestation in some settings. The following sections of this chapter use the assembled case studies to explain these patterns of change in south Asian forests.

Forests in the Colonial and Immediate Postcolonial Eras

Historical evidence about early patterns of forest cover change exists for only some regions, one of which is West Bengal. When the British first began to assert control over West Bengal during the late nineteenth century, they found a largely agricultural landscape with established villages in the plains. Forests covered the hills to the southeast of the Ganges plain. The forested hill districts contained small populations of shifting cultivators. Throughout the nineteenth century, hunters reduced the dangers from wild animals such as tigers and elephants through programs of "vermin eradication," and farmers pushed the margins of permanently cleared land into the formerly forested hill districts (Sivaramakrishnan,

Table 8.1
Forest-Cover Change in South Asia: 1980s and 1990s

Nation	1980–1990 (%)	1990–2000 (%)
Bangladesh	–3.9	+1.3
India	–0.6	+0.1
Nepal	–1.0	–1.8
Sri Lanka	–1.3	–1.6

SOURCES: Food and Agricultural Organization (FAO). 1993. "Forest Resources Assessment, 1990—Tropical Countries." FAO Forestry Paper, #112. Rome; and FAO. 2001. "Forest Resources Assessment, 2000." FAO Forestry Paper #140, Rome.

1999:34–120). In the more remote parts of the subcontinent such as the central Himalayan districts in northern India, the expansion of settlements and agriculture into blocks of forest continued into the first part of the twentieth century (Rathore et al., 1997:268).

British colonial administrators grabbed resources in nineteenth-century India much as they did in twentieth-century East Africa. In the 1850s and 1860s, the colonial regime embarked on an ambitious program of railroad expansion, and administrators quickly became concerned that local forests did not contain an adequate supply of wood for the sleepers under the railroad tracks. At one point during this period, the British imported the wood for sleepers from Europe. To secure a local source of supply, British colonial administrators authorized the creation of reserved or "closed" forests, beginning in 1878. Under the new law, peasants could no longer gather minor forest products, graze their livestock, or cultivate lands within the newly designated reserves without obtaining permits and paying taxes to the authorities (Sivaramakrishnan, 1999:192). These rules deprived Indian peasants of an important source of subsistence and violated established traditions of local governance (Guha, 2000:36–39). In The Dangs, a hilly region in southern Gujarat, foresters created the reserves in increments, incorporating lands with few shifting cultivators in a particular year into the reserves and leaving cultivated lands outside of the reserves (Hardiman, 1996:122–123). Through this process, colonial officials gradually appropriated a large proportion of the less densely populated lands in The Dangs into reserves.

The new order violated the "moral economy" constructed by villagers around Indian forests, and they responded with both overt and covert acts of resistance (Scott, 1975, 1985; Guha, 2000:62–151).[1] When the newly independent state of India retained the centralized system of reserved forests, the acts of peasant resistance continued, culminating in the celebrated Chipko movement during 1970s when women from Himalayan villages "hugged" trees to protect them from commercial loggers. The centralized system of control also worked against the emergence of conservation practices among villagers. Conservation without control made no sense. The most dramatic instances of peasant conservation occurred, as in the Chipko case, where peasants retained control over the forests despite challenges from outside interests.

After independence, state officials continued to control the forests. They justified their control as part of an effort to accelerate economic development in the larger society. International development agencies assisted the state in this endeavor. For example, the World Bank provided startup funds during the 1970s for an ambitious scheme to replace imported timber with plantation timber grown on state lands in the Bastar district of Madhya Pradesh. The project's administrators planned to destroy the sal forests native to the Bastar region and replace them with plantations of Caribbean pine *(Pinus caribaea)* that would supply new, technologically advanced paper mills. The project's planners never seriously considered this scheme's impact on the tribal people who relied on the sal forests for a major portion of their sustenance. Faced with protests by the tribals, falling prices for paper, and a reluctance of private companies to invest in the project, Indian forestry officials decided to abandon the Bastar project in the early 1980s (Anderson and Huber, 1988). The hostility between Bastar villagers living in the forests and foresters policing its use typified the sentiments of people in most other forested regions of the Indian subcontinent between 1950 and 1980.

The meta-analysis, reported in part A of table 8.2, outlines the forest cover dynamics that emerged during this period. Most of the case studies in both the earlier and later periods were carried out in northern India or Nepal. Loggers, those with and those without contracts from forestry departments, cut stands of timber in forest reserves, while growing populations of shifting cultivators put increasing pressure on the remaining accessible forests, both for fields and for fuel (Bajracharya, 1983). To accommodate the increased demands on the land, cultivators reduced the length of their fallows, leading to less robust secondary growth in the fallowed fields (Faminow and Klein, 2000). In some instances, settlers would move into the state forests, and the government would, after a period of conflict, acknowledge the invaders' claims to land. By regularizing

Table 8.2
Deforestation in South Asia During the 1980s and 1990s:
A Qualitative Comparative Analysis

A. Pre-1980s and 1980s (12/12)
FUELWOOD POP central (4) +
LOGGING smallag FUELWOOD central (1) +
LOGGING SMALLAG fuelwood pop CENTRAL (1) +
logging SMALLAG fuelwood pop central (2) +
LOGGING SMALLAG FUELWOOD POP (3)

B. 1990s (12/14)
SMALLAG FUELWOOD central pop (1) +
SMALLAG FUELWOOD central logging (3)

The "+" in this table indicates that deforestation occurs whenever any of the listed combinations of causal conditions exists.

Upper case indicates that the causal factor is present, and lower case indicates its absence. The ratio in parentheses indicates the proportion of studies for that period whose findings agree with the Boolean expressions in the lines below it. The number in parentheses after each causal combination indicates the number of studies whose findings fit that causal conjuncture.

Factors: CENTRAL = centralized control of forests; FUELWOOD = changes in firewood collection; LOGGING = commercial forestry enterprises; POP = population growth; SMALLAG = small-scale agriculture.

these claims to land, government officials encouraged other encroachments. In some districts, a series of encroachments led to rapid declines in forest cover. One district in Andhra Pradesh lost 45 percent of its forest area between 1983 and 1993 (Mukherjee, 1997).

Growing urban populations consumed increasing amounts of fuelwood, so the declines in forest cover occurred particularly rapidly around Indian cities (Bowonder, Prasad, and Unni, 1987). Figure 8.2 shows the ovens that transform fuelwood into charcoal for sale in urban areas. Commercial interests also logged state forests, both legally and illegally, to obtain poles and wood for eventual sale to builders in cities (Rawat, 1995). Under these circumstances, forest cover continued to decline in India and Bangladesh until the 1980s. In some places, as indicated in part B of table 8.2, this dynamic continued to cause losses of forest in the 1990s.[2]

Beginning in the 1970s, the rising prices for forest products in largely deforested areas induced villagers to allow degraded forest to regenerate (Foster and Rosenzweig, 2003). Government officials had begun trying to regenerate destroyed forests much earlier, starting in 1794 when the British began to encourage the planting of teak in Bengal. The colonial regime's initiatives did not lead to widespread tree planting (Sivaramakrishnan,

Figure 8.2   Ovens for making charcoal near Dehra Dun, India. Credit:
G. Bizzarri, FAO photo (1996).

1999:107–111). During the 1960s and 1970s, government officials again attempted to replenish the stock of standing wood by establishing tree plantations, but this program made only marginal gains because few of the seedlings in the plantations survived. Survival rates on government plantations in Uttar Pradesh during this period ranged from 7.6 to 30 percent of the seedlings planted. Initiatives by elders in some communities to afforest degraded and denuded areas, sometimes with government assistance, increased both tree cover and biodiversity, but these efforts remained isolated experiments, so they had little impact on overall forest cover trends (Guha, Prasad, and Gadgil, 1984).

Whereas elites in centralizing governments shaped the fate of Indian forests through coercive policies between 1850 and 1980, variations from this larger historical pattern emerged in other places. In Sri Lanka, the declines in forest cover occurred early in the colonial period when British planters carved coffee and tea plantations out of upland forests during the second half of the nineteenth century (Wickramagamage, 1998). In Nepal, the eradication of malaria during the mid-twentieth century in the Terai district at the base of the Himalaya encouraged both spontaneous and government-assisted settlement and deforestation of the zone (Soussan, Shrestha, and Uprety, 1995). In these instances, as in the more widespread pattern described earlier, officials and citizens of an expanding nation-state, working in some instances for a global imperial enterprise, found that destroying forests furthered their common and more personal purposes.

Although this pattern of exploitation continued in some places, the destruction of the forests eventually brought changes. During the later part of the twentieth century, the prices for wood remained high, but logging firms lost interest in the small remnant forests in many parts of India and Bangladesh because they contained such small stocks of commercially valuable timber. By 1980, many logging companies could not find stands of wood to exploit, and forest-based industries faced the threat of closure (Bhat, Murali, and Ravindranath, 2001:607). Under these circumstances, logging companies became less important in driving deforestation in south Asia. They appear in the early case studies (part A) and disappear in the later case studies (part B) summarized in table 8.2.

Forest-Cover Changes in the 1990s: Joint Forest Management and Plantations

The turnaround in forest-cover trends on the Indian subcontinent had its origins in earlier processes of degradation and destruction. In response to the growing scarcity of wood in many cutover and degraded forests, the real prices for fuelwood began to rise during the 1970s (Foster and Rosenzweig, 2003:608). The rise in prices encouraged villagers to view regenerating natural forests as a potentially significant source of future income. At the same time, forestry officials acknowledged that their efforts to establish tree plantations could not keep up with the large amount of natural forest lost each year through degradation and deforestation (Poffenberger, McGean, and Khare, 1996:21).

Beginning about 1980, both villagers and foresters began trying to increase the supply of wood. In the poor eastern state of Orissa, villagers organized to protect regenerating sal forests after extensive cutting had destroyed almost all of the forests in their districts. When the first local efforts to protect forests showed signs of succeeding, neighboring villages quickly established their own forest protection committees, and geographic clusters of protective efforts emerged in all but the most forest rich parts of the provincial interior. By July 2001, approximately 5,000 Orissa villages had begun to manage local forests, and 1,200 of the villages had entered into formal co-management agreements with the provincial forest agency (Singh, 2002:36).

Federal foresters and politicians expedited the changes in communities through changes in state forest policy. The policy changes began with India's Forest Conservation Act of 1980, which banned the conversion of forests for agricultural and other purposes. State-sanctioned clearing declined dramatically after 1980, but illegal encroachments into state forests continued. The illegal logging, the encroachments, and the growing expanse of degenerated, cutover sal forests in densely settled areas gradually convinced foresters and politicians that forestry departments could not protect the state's forests. This realization provided the political impetus for legislation enacted in 1988 and 1990 that authorized Indian states to create Joint Forest Management (JFM) programs between local communities and forestry departments (Poffenberger and Singh, 1996:62).

Joint Forest Management does not cede definitive control over forests to local leaders; in this sense, it represents a power-sharing agreement that involves the transfer of some but not all control over forests to villages (Hussain, Bibi, and Kaushal, 1999). By giving locals a share of the benefits flowing from the forests, the reform altered the political stance of local leaders toward the state's forest managers. By excluding peasants from the forests, the colonial institutional order encouraged peasants to focus on how they might gain access to the forests. The health of the forests, what local residents might find when they gained access, remained a secondary concern. By promising a stream of benefits flowing from the forests to villagers, the new institutional order encouraged villagers to concern themselves with the natural regeneration of degraded or cutover forests. Under these changed circumstances, villagers began to express conservation concerns and criticize state forestry officials for lax enforcement of forest protection rules (Agrawal, 2001).

In Andhra Pradesh, in southern India, villages in close proximity to degraded forests formed local forest protection committees, with a mandate to include village women on each committee. Nongovernmental organizations frequently served as go-betweens, providing lines of communication between village committees and the forestry departments. As members of committees, villagers received rights to gather nontimber forest products and to cut small timber as part of silvicultural treatments. Through JFM, provincial forestry agencies employed villagers to plant trees and maintain the forests. Village members could also expect a portion of the proceeds from the sale of wood from the forest. The creation of the committees had several immediate effects. Illegal logging ended. Villagers evicted encroachers from outside the region from the forests, and no more encroachments took place. Villagers also began to suppress forest fires, thereby allowing both secondary forests and brush to thicken and spread. The more luxuriant growth made it possible for villagers to harvest much more grass and twigs from the forests than they had in previous years (Mukherjee, 1997).

Studies in other locales report similar patterns of stability or recovery. During the 1990s, the state forestry department in Madhya Pradesh initiated a large program of joint forest management with support from the World Bank (Stone and D'Andrea, 2001:87–95). Tree-planting initiatives spread rapidly from village to village in Nagaland, in northeastern India (Faminow and Klein, 2000). Orissa did not begin to facilitate JFM until the late 1990s, but villagers there did not wait for official approval. With modest sums of money from the Oxfam India Trust, they formed forest protection committees and an association of the committees to defend biodiversity in the region (Stone and D'Andrea, 2001:104–114). Nepal also initiated a community forestry program. In the Terai region alone, villagers had formed more than 320 forest user groups by 1997 (Poharel, Adhikari, and Thapa, 1999:27). Studies of forest-cover change in the middle hills region of Nepal report increases in forest cover during the 1990s, and they tie the increases to a strengthening of local governance over forests during the 1980s and 1990s (Fox, 1993, Schreier et al., 1994; Jackson et al., 1998, Varughese, 2000).

People also planted trees outside community managed forests. Smallholders in Bangladesh began creating woodlots on their farms, practicing "homestead forestry." Farmers with larger tracts of land and nonfarm incomes established more woodlots. High prices for wood accelerated the pace of woodlot planting (Salam, Noguchi, and Koike, 2000). In the 1980s, the Indian government launched a large-scale afforestation program that subsidized the creation of short rotation tree plantations on smallholder and village lands. A relatively large proportion of the seedlings, 77 percent, survived the 6 to 8 years until they were harvested for sale. The program has benefited farmers with commercial interests, but restrictions on who could harvest products from the new forests prevented the very poor from taking advantage of the new sources of fuelwood and fodder (Bhat, Murali, and Ravindranath, 2001:611).

Table 8.3 presents a qualitative comparative analysis of the human factors associated with afforestation. The devolution of forest policy to the village level through programs such as JFM occurs in five of the six causal conjunctures in the table. Devolution became almost a necessary condition for increases in forest cover in south Asia. In valleys in the mid-hills of Nepal, where outside agencies failed to promote community forestry, forest cover continued to decline during the 1980s and 1990s (Schweik Adhikari, and Pandit, 1997). Declines in the human pressure on the forests stemming from the creation of nonfarm livelihoods may also explain why peasants do not feel a need to expand agriculture at the expense of forests around villages (Bluffstone, 1995; Jackson et al., 1998). Sher Bishwakarma's story (see box) illustrates how individual decisions to move into a nonfarm occupation, made in response to changing environmental and economic conditions, can reduce the pressures that humans put on south Asian forests.

Data on tree plantations from the FAO's Forest Resource Assessment 2000 provides an additional perspective on the turnaround in forest cover trends in south Asia. Although China has more forest plantations than any other nation in the world, the annual planting rate in India, 1,509,000 hectares, was the highest in the world during the 1990s.[3] Plantations provided 50.8 percent of the tree cover in India and 46.8 percent of the tree cover in Bangladesh in 2000. These are the world's highest rates of planted woodlands outside of the British Isles.[4] In contrast, plantations constituted only 9.4 percent of the tree cover in Indonesia in 2000 (FAO, 2001).

Although the widespread planting of trees provided poor rural residents with important resources and sequestered carbon, tree plantations usually added little to the biodiversity of a region. For example, Pine *(Pinus roxburghii)*-dominated tree plantations in watersheds close to Kathmandu, Nepal, prevent soil erosion and provide timber for construction, but they have few other uses. The blanket of pine needles that covers the forest floor in these plantations cannot be used as animal fodder; it does not burn well; it suppresses other types of vegetation, and it creates an acidic soil (Schreier et al., 1994). Eucalyptus *(Eucalyptus rostrata),* the other exotic featured in many tropical plantations, has a similar set of characteristics, so it also can impoverish plant biodiversity in a place. Plantations may indirectly benefit biodiversity in a region because, where local people can draw on plantations for forest products, they take fewer

Table 8.3
Afforestation in South Asia in the 1990s:
A Qualitative Comparative Analysis

1990s (12/14)

   logging SMALLAG pop LOCAL (2) +
   pop SMALLAG fuelwood LOCAL (2) +
   LOGGING smallag pop LOCAL (1) +
   LOGGING fuelwood pop LOCAL (1) +
   logging smallag FUELWOOD POP LOCAL (1) +
   SMALLAG FUELWOOD POP LOGGING local (1)

The "+" in this table indicates that afforestation occurs whenever any of the listed combinations of causal conditions exists.

Upper case indicates that the causal factor is present, and lower case indicates its absence. The ratio in parentheses indicates the proportion of studies whose findings agree with the Boolean expressions in the lines below it. The number in parentheses after each causal combination indicates the number of studies whose findings fit that causal conjuncture.

Factors: FUELWOOD = changes in firewood collection; LOCAL = significant local participation in control of forests; LOGGING = commercial forestry enterprises; POP = population growth; SMALLAG = small-scale agriculture.

forest products from the more biodiverse, naturally occurring forests in a region (Kohlin and Parks, 2001).

Policy changes, coupled with prior losses in forest cover, best explain the recent change in south Asian forest cover trends. Several other historical conditions facilitated the change. The region had a long tradition of local community governance, dating back to the pre-colonial era, so the routines of local governance were familiar to many villagers. Rural population densities in south Asia have been and continue to be the highest of all of the tropical regions under examination. To some extent, the high densities have facilitated the management of forest resources because each village has contained enough people to monitor and enforce local rules about the use of the relatively small forests nearby (Gibson, McKean, and Ostrom, 2000:235). Under these circumstances, the devolution of power over forests resulted in the rapid creation of effective community institutions for governing the use of forests in many places.

Spatial Variations in Afforestation and Deforestation in South Asia

Although people have begun planting trees in a growing range of places in south Asia during the past 15 years, the changes in forest cover have occurred in spatially uneven ways. People afforested in some areas, but

---

Sher Bishwakarma's Passage from the Forest as a Child to the Schoolhouse as an Adult

When his 13-day-old first son joined the human community through the nwarani ritual in 2000, Sher Bishwakarma talked about his hopes for his son, Dil. "My deepest wish for my son is that he will get a good education. When I was a child, we did not get the opportunity to go to school. Instead we had to look after cattle, cut leaves, and work in the fields. Today, it is difficult to make a living up in the hill villages, so, if my son wants to go to Kathmandu or India when he grows up, he will need some knowledge."

Bishwakarma, a member of the Dalit caste of untouchables, lives in Pachnali village in western Nepal. His parents and grandparents, following the dictates of the caste system, worked as blacksmiths in the village. They made farm implements and kitchen utensils. To do so, they would first make charcoal from wood gathered in nearby forests. Human pressures on these forests eventually destroyed them, so now blacksmiths from Pachnali must collect wood in distant forests. These difficulties and the presence of other blacksmiths in Pachnali have persuaded Bishwakarma that he should try another occupation. He has had 5 years of schooling, so, rather than working in the fields or forests, Bishwakarma now teaches illiterate women to read during the evenings. He hopes that his newborn son will follow his own occupational path from the forests to a schoolhouse or office.

SOURCE: Madsen (2003).

---

they deforested in other areas; one watershed experienced afforestation, but another one did not. A suggestive study of forest cover dynamics in the middle hills of Nepal between 1978 and 1992 established that forest cover had increased in the valley bottoms close to roads where private landowners had planted more trees on their own lands, and communities had established tree plantations on scrub and pasture lands. Degradation occurred on the upper slopes of the hills farther from roads and villages. People overgrazed pasture lands, stripped the forests of fuelwood, and allowed fires to burn out of control. Peasants also abandoned some infertile fields on the upper slopes to devote more time to off-farm occupations (Jackson et al., 1998). Interest group politics appeared to reinforce the center–periphery pattern in forest-cover change. Commercial logging firms had few interests in the degraded and deforested areas near villages where JFM efforts to regenerate forests have concentrated. In more forested districts, which are usually more remote, logging firms want to retain logging concessions, and they oppose JFM schemes in these locales (Poffenberger and Banerjee, 1996:324; Poffenberger, McGean, and Khare, 1996:38)

In another part of the middle hills of Nepal, the level of collective action interacted in complex ways with the distribution of land and people in a place. Communities in remote, off-road locations often had difficulty monitoring their forests, especially in circumstances where divisions of caste and class weakened collective efforts. Inattention by district officials to these communities sometimes played a role in the decline of their forests (Varughese, 2000:219–220). To the extent that communities located in favored, lower-elevation locations closer to roads are more populous than communities located in less accessible places, the size of a community's population may affect its ability to establish and carry out a program to increase forest cover in a region. Communities with larger populations sometimes manage forests more effectively than communities with smaller populations because forest protection committees in the larger communities have enough members to monitor their forests and carry out forest improvement projects (Agrawal, 2000).[5]

Taken together, these findings suggest a center–periphery pattern in institutionally driven patterns of forest recovery. Forests prosper in communities close to roads where villages have large forest protection committees that often receive outside assistance. Forests disappear or undergo degradation in remote settings where communities have difficulty regulating and managing the use of forests. In these places, open-access property regimes persist, and peasants scramble to appropriate forest resources for themselves. This pattern may explain some of the discrepant findings about forest-cover change in south Asia. Windshield surveys and treks by professionals working in Nepal indicate substantial forest recovery in the middle hills of Nepal, whereas remote sensing analyses of forest-cover change in Nepal for the late 1980s and early 1990s indicate continued declines in forest cover (Ortiz-Chour, 1999).

The different assessments of forest-cover change reflect the different universes in which the analysts work. The national and expatriate forest professionals work primarily with the more accessible communities and see the forests closest to these communities where, indeed, forest cover has recently increased. The remote sensing analyses include both the more and the less accessible of the forested regions. Because the less accessible areas constitute a larger area, net deforestation still characterized Nepal during the 1990s. Differences in tenure as well as in location distinguish the recovering from the declining forests. Open access continues to characterize state forests, whereas community-managed forests enjoy higher levels of protection (Kanel and Shrestha, 2001). Communities tend to manage those forests closest to them, so state forests tend to be located in more peripheral places where it is more difficult to restrict access to the forests.

In this regard, it is worth remembering that communities managed only about 3 percent of India's protected forests and 11 percent of Nepal's protected forests in the late 1990s (Stone and D'Andrea, 2001:224; Kanel and Shrestha, 2001:699). Given the small area under community forest management, the recent turnaround in forest-cover trends in India must be attributable to accelerated schedules of tree planting by state forestry departments, to expansion of small forests on farms, and perhaps to spontaneous regrowth on lands that peasant households with off-farm employment now neglect. These surmises remind us of the conjunctural, multicausal nature of changes in forest cover trends.

The center–periphery pattern of environmental change described here recalls patterns in China. A recent report on environmental management in villages of north China (Lee, 1999) describes well-tended fields and abundant tree planting close to villages, and transitory, exploitative use of eroded lands farther from village centers. The difficulties of transporting tools and manure to the peripheral fields discouraged farmers from undertaking environmental improvements on outlying lands. Netting (1993) has observed a somewhat similar pattern of environmental change at a larger, regional scale in China. In the densely populated, long-inhabited rice-producing region along both sides of the Yangtze River, small-scale, sustainable agriculture emerged in the midst of a considerable population increase after the seventeenth century. At the same time, in the more peripheral regions of northern China along the Huang-Ho, much smaller and more transitory populations of cultivators contributed to severe problems of soil erosion. In this setting, as in south Asia, remoteness does not confer protection on forests the way that it does in South America or central Africa. Instead, a peripheral location contributes to a weak institutional order that makes unbridled exploitation of the forests more likely.

Conclusion

These circumstances pose special challenges for people concerned with forests in south Asia. As the quote that opens this chapter makes clear, rural populations in south Asia have for centuries lived at much higher densities than rural populations in other tropical regions. For this reason, they have degraded or destroyed remote forests to a greater degree than elsewhere. Recent trends in homestead and community forestry promise to restore some forests, but the larger forested domain, often in more sparsely populated districts, remains in the hands of state foresters. The way they manage these forests will have an important impact on the overall health of south Asian forests.

Although initial efforts to decentralize natural resource regulation have yielded promising results, the reforms have also presented new challenges to participants in sustainable development efforts. First, the devolution of political power to local jurisdictions during the 1990s increased the ability of local politicians to control the appointments of guards in national parks. In at least one of India's national parks, Nagarahole in Western Ghats, the increased power of local politicians made it more difficult for park managers to preserve natural resources. Obliged to supporters in communities surrounding the park, local politicians provided political protection for their supporters who exploited natural resources inside the parks (Karanth, 2002:193). Second, the leaders of community forestry organizations formed umbrella organizations in the mid-1990s that allowed some community leaders to speak for many communities when they negotiated with state or federal forest bureaucrats. As the power of community forest organizations grew, both separately and through coalitions, local forest politics intensified, and in the resulting struggles for control, village elders sometimes marginalized women and other historically oppressed groups (Sarin, 2001; Singh, 2002). As a result, issues of representation within these organizations have become more pressing. Third, the devolution of control over forests to local governments has challenged them to do an effective job of managing a newly acquired common property in a context where free riding and preferential access can occur. Forest user groups among the local inhabitants offer the most potential as stewards for the newly acquired forests (Blair, 1996). Finally, where spontaneous generation occurs, it produces more biodiversity, but the stream of income may not match the income coming from a plantation (Lugo, 1992). Tropical tree plantations have promised substantial returns, but their record of production has been quite disappointing, with actual yields averaging only 20 to 30 percent of the anticipated yields (Persson, 1996). Clean Development Mechanism payments authorized by the Kyoto treaty have recently provided another incentive to afforest by promising villagers another potential source of income from forest planting (Poffenberger et al., 2002).

Southeast Asia: Deforesting the Lowlands, Afforesting the Highlands

The vegetation was most luxuriant, comprising enormous forest trees, as well as a variety of ferns, caladiums, and other undergrowth, and an abundance of climbing rattan palms.
ALFRED RUSSEL WALLACE, *The Malay Archipelago* (1869:19)

Wallace's eyes did not deceive him. The tropical rain forests of Southeast Asia contain the world's most diverse assemblage of vascular plants as well as its most economically valuable hardwoods. Because these forests stretch across an extensive archipelago from the Solomon Islands in the Pacific to the upland regions of continental Southeast Asia, they are more fragmented than the other large blocks of rain forest, in the Amazon and central Africa. The insular setting of the rain forests also ensures that, until recently, a fairly large proportion of them existed near coasts. The coastal locations of the forests did not make them easy to exploit. Moving inland in Southeast Asia has usually meant moving uphill. Limited coastal plains and extensive uplands characterize Indochina on the mainland. Mountainous terrain, topped by Kinabalu at 4,101 m in Kalimantan and Puncak Jaya at 5,030 m in New Guinea, exists in the interior of almost all of the islands. A growing proportion of Southeast Asian rain forests exist in these hard-to-exploit, montane or submontane settings (Whitmore, 1995:6). Figure 9.1 maps the main features of Southeast Asian rain forests. For our purposes, the region includes Myanmar, Thailand, Laos, Vietnam, Cambodia, Malaysia, Indonesia, Philippines, and Papua New Guinea.[1]

In addition to mangrove and upland forests, the region contains extensive peat and freshwater swamp forests in eastern Sumatra and southern Irian Jaya. Tropical monsoon forests characterize places with dry seasons, such as the rain-shadowed areas east and west of the Irrawaddy River valley in Myanmar. When logging and farming disturbs or destroys these forests, a degraded forest, dominated by invasive species, often replaces it. For example, degraded, bamboo-dominated forests are particularly extensive in the Arakan Yomas region of Myanmar. In Laos, monsoon forests have replaced evergreen rain forests after disturbance (Collins, Sayer, and Whitmore, 1991:106, 166). As Figure 9.1 indicates, degraded forests are particularly widespread in the upland areas of mainland Southeast Asia.

Political turmoil has unsettled daily life in Southeast Asian societies since World War II. Insurrections broke out when Europeans tried to reoccupy their prewar colonial dominions after the war, and the unrest continued on a lesser scale after independence. During the 1950s and 1960s, every newly independent state in the region engaged in armed conflict with insurgents in rural settings. Civil conflict declined after 1970, but states remained weak throughout the region. The legitimacy of even the most democratic regimes remained suspect because government bureaucrats used networks of patron–client ties to give wealthy families preferential access to government-controlled resources. These practices made it difficult to implement economic or environmental regulations.

The leaders of these weak regimes governed rapidly changing societies. Populations grew rapidly after World War II. Between 1980 and 1999, Indonesia's population increased by 40 percent, from 148 million to 207 million people. Since 1980, fertility rates have begun to decline in the more affluent counties in the region such as Thailand, but annual rates of population growth still remain quite high, ranging from 1.3 percent per year in Thailand to 2.9 percent per year in Cambodia (World Bank, 2001). Unlike most other countries in the tropics, several Southeast Asian nations, Malaysia, Thailand, and Indonesia, experienced vigorous economic growth after 1980. Although economic growth rates were

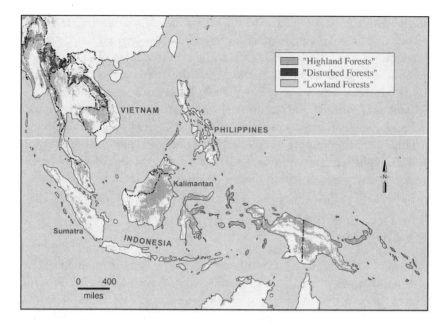

Figure 9.1 The forests of Southeast Asia.

impressive from 1980 to 1997, people were extremely poor at the outset of the period and remain only moderately better off at present. Economic prosperity has transformed Malaysia (1999 gross national product per capita, U.S.$3,390) and Thailand (1999 GNP per capita, U.S.$2,010) into middle-income developing countries, but in other countries such as Indonesia (1999 GNP per capita, U.S.$600) and Philippines (1999 GNP per capita, U.S.$1,050) recent economic growth has brought prosperity only to urban residents. Most inhabitants of rural areas in these countries continue to earn very low incomes. Differences between rural and urban areas in incomes and in the provision of basic services spurred rural-to-urban migration during the late twentieth century. Between 1980 and 1999, the cities' share of the regional population increased from 23.6 to 35.9 percent (World Bank, 2001).

Over the last two decades of the twentieth century, people destroyed Southeast Asian rain forests more rapidly than they destroyed large blocks of forest in central Africa and Latin America. Between 1990 and 2000, forest cover declined at 0.9 percent per annum in the Southeast Asian region, compared with declines of 0.5 percent and 0.3 percent, respectively, in the Amazon and central African basins during the same time period (FAO, 2001). A similar regional differential characterized deforestation rates during the 1980s (FAO, 1993). The reported rates of deforestation in Southeast Asia increased slightly from the 1980s to the 1990s in the two countries with the largest remaining blocks of forest, Indonesia and Myanmar. All of the other countries, with smaller forests, reported lower rates of decline in forest cover during the 1990s than they did during the 1980s.[2] In the remainder of this chapter, I try to account for both the relatively rapid regional rates of deforestation and the increasing diversity of forest-cover trends within the region during the 1990s.

The Destruction of Southeast Asian Forests from 1650 to 1990

The rain forest regions of Southeast Asia have long contained extensive populations of small-scale farmers who cultivated wet rice. In some places, these activities may have led to permanent deforestation, but in other places forests disappeared and then reappeared in altered form when shifting cultivators moved elsewhere. This cycle of destruction and regeneration changed in the late seventeenth century when Europeans began to exploit the naturally occurring teak forests of Java. By the middle of the nineteenth century, Dutch merchants and colonial officials had depleted all of the coastal and some of the interior teak forests on Java (Peluso, 1992:27–78). To ensure a continued stream of revenue from the sale of teak, nineteenth-century colonial administrators established state forests, introduced ideas of "scientific forestry," and began to regulate timber harvesting in Java.

The destruction of tropical rain forests spread to the mainland and to other islands in the Malay Archipelago during the nineteenth and early twentieth centuries. Foreign investors favored by the colonial regimes began to convert coastal forests into plantations that produced crops for sale in recently created global markets. In the Irrawaddy delta of Burma, British colonial administrators, interested in promoting rice exports, encouraged Burmese smallholders to convert forests into rice paddies (Adas, 1998). Dutch investors established tobacco plantations in eastern Sumatra (Pelzer, 1968). In southern Philippines, American, Japanese, and European landowners created coconut palm plantations (Simkins and Wernstedt, 1971). To accommodate populations that began to increase in the early twentieth century and to consolidate colonial power in peripheral regions, Dutch and American administrators created new land-settlement schemes in Sumatra (Dutch) and Mindanao (American) in the two decades prior to World War II (Pelzer, 1945).

Following independence after World War II, the configuration of forces driving deforestation changed somewhat, with loggers, colonization program officials, and heads of smallholder households working together to convert forests into fields. The leaders of the newly independent states decided to continue the prewar colonization programs. By the 1970s, Indonesia, Malaysia, Philippines, and Thailand all had established colonization programs that converted tracts of tropical rain forest into agricultural communities (Uhlig, 1988). By the late 1980s, the Indonesian transmigration program, the largest colonization program in the region, had resettled over 5 million migrants in Indonesia's outer islands (Holden and Huoslef, 1995). The new settlements and related roads extended the state's authority into rural areas and quelled political unrest in the sending regions through "land to the poor" campaigns.

For every government-sponsored migrant who settled in a planned village, five or six migrants settled spontaneously in forested interior or upland locations in the region. Usually, improvements in infrastructure figured in the spontaneous migrants' decisions to move to a forested place and start a farm (Amacher et al., 1998). The new roads made it easier to get to the land and provided a means for marketing crops from the farms. If a settler laid claim to a tract of land in advance of road building, he could reap speculative gains from the increase in land values that occurred when the construction crews completed the segment of the road closest to his farm.

Often, logging firms built the roads that spontaneous colonists used to gain access to an area (Vayda and Sahur, 1985). The loggers would build the roads to extract the timber, and migrants would then establish smallholdings along the roads. This sequence of loggers followed by settlers had an additional important advantage for colonists. In Southeast Asia, as elsewhere in the tropics, loggers take trees selectively, harvesting the commercially valuable species and leaving other trees standing. In much of Southeast Asia, and especially in Sumatra, Kalimantan, Philippines, and mainland Southeast Asia, the commercially valuable trees of the genus *Dipterocarpaceae* dominate stands, and, as a result, loggers remove a much higher proportion of trees from stands in Southeast Asia than they do elsewhere in the tropics.[3] By removing the trees, the loggers reduce the amount of work that migrants have to do in preparing the land for cultivation.

Table 9.1

Deforestation in Southeast Asia: A Qualitative Comparative Analysis

---

A. 1980s (27/33)

    LOGGING SMSEDAG pop (1) +
    LOGGING SMSEDAG hills (2) +
    LOGGING SMSEDAG colon (7) +
    SMSEDAG pop hills (4) +
    SMSEDAG colon hills (1) +
    LOGGING POP COLON hills (4) +
    logging POP COLON HILLS (3)

B. 1990s (21/21)

    LOGGING PLANT intense TENURE (3)
    SMSEDAG intense tenure roads (2) +
    LOGGING smsedag intense TENURE ROADS (2) +
    LOGGING SMSEDAG PLANT intense roads (3) +
    logging SMSEDAG PLANT intense tenure (1) +
    logging SMSEDAG plant intense roads (2) +
    LOGGING SMSEDAG plant INTENSE TENURE roads (1) +
    logging smsedag plant intense tenure ROADS (1)

The "+" in this table indicates that deforestation occurs whenever any of the listed combinations of causal conditions exists.

Upper case indicates that the causal factor is present, and lower case indicates its absence. The ratio in parentheses indicates the proportion of studies for that period whose findings agree with the Boolean expressions in the lines below it. The number in parentheses after each causal combination indicates the number of studies whose findings fit that causal conjuncture.

Factors: COLON = government colonization programs; HILLS = marginal terrain, ill suited for agriculture; INTENSE = agricultural intensification occurring; LOGGING = logging; PLANT = plantation expansion; POP = rural population increase; ROADS = construction of roads into rain forest region; SMSEDAG = small-scale, sedentary agriculture; TENURE = conflicts over land tenure.

Part A of table 9.1 outlines a qualitative comparative analysis of 33 studies of tropical deforestation in Southeast Asia from the end of World War II until the end of the 1980s. All of the major countries in the region, with the exception of Myanmar and Papua New Guinea, are represented in the data set. The analysis largely confirms the configuration of driving forces outlined here. Logging companies, colonization agencies, and publicly funded road-building programs created circumstances that rewarded the conversion of forests into fields. Smallholders took advantage of these opportunities and invested their labor in the creation of new farms. In lowland areas well served by roads, settlers converted the forests into fields, where they practiced continuous cultivation (Cropper, Puri, and Griffiths, 2001; Fox et al., 1995; Liu, Iverson, and Brown, 1993).

Early analyses of tropical deforestation identified growing populations of shifting cultivators as the driving force behind deforestation in Southeast Asia (Myers, 1984). To quote one representative source (U.S. Department of State, "Comments on Biological Diversity Action Plan," Cable 23811, quoted in Brady, 1988:410), "In Thailand the principal threat to long-term maintenance of biological diversity and tropical forest resources is agricultural encroachment on already designated conservation areas by nomadic hill tribes and landless lowland Thais."

To ferret out the contributions of swidden cultivators to deforestation, I separated them from small-scale, sedentary cultivators in coding the studies. Despite the early assertions by Myers (1984) that shifting cultivators played an important role in Southeast Asian deforestation, the meta-analysis suggests that they made a relatively small contribution to the overall decline in forest cover. For that reason, SWIDDEN does not appear in the causal configurations in part A of table 9.1. Although some swidden cultivators, such as the Hmong of northern Thailand, cleared considerable amounts of rain forest (Geddes, 1976; Kunstadter and Chapman, 1978), others, such as the Kenyah Dayaks of interior Borneo, converted little forest into permanent fields (Colfer, Peluso, and Chung, 1997). During the same period, migrants from the lowlands, practicing small-scale, sedentary agriculture, did deforest extensive areas in upland areas of Philippines and Thailand (Kummer, 1992; Cruz, 2000; de Lang, 2002).

Although the emphasis on shifting cultivators as a force driving deforestation in Southeast Asia has been justifiably criticized for ignoring other factors (Lambin et al., 2001), it may have been historically accurate for the 1950 to 1980 period in one general way. Rural populations did expand rapidly during this period and urban labor markets remained too small to provide work for most young adults. In this context, young people forming households in rural areas often looked at the uncultivated margins of the rain forest as a place where they could invest their labor in an enterprise that would yield a stream of benefits into the future. Carving a farm out of the forest would seem like a wise investment when other highly capitalized actors such as a government colonization agency or a logging firm build the roads that provide smallholders with a means for marketing crops from their farms. The interaction between growing populations of smallholders and organizations dedicated to opening up peripheral regions for exploitation best explains the first wave of deforestation in Southeast Asia after World War II.

Forest Destruction in the 1990s

During the 1980s, the configuration of forests driving tropical deforestation in Southeast Asia began to change in fundamental ways. The
decline in rural insurgencies and the outcry over the destruction of
tropical rain forests reduced the support for colonization programs,
so by the late 1980s they had ceased to be a major contributing factor
behind deforestation in the region.[4] With the continuing urbanization
of Southeast Asian societies, labor markets in cities began to pull significant numbers of smallholders off the land, especially in the more
prosperous nations. New land settlement schemes in peninsular Malaysia lost many of their smallholders to urban labor markets in the
1980s and 1990s. With farm labor becoming difficult to obtain, smallholders in peninsular Malaysia allowed some of their marginal lands to
revert to forest. Malaysian oil palm entrepreneurs decided to expand
their plantations in neighboring countries such as Indonesia, where
farm workers were plentiful and less expensive to employ (Brookfield,
1994; Potter and Lee, 1998). Loggers also shifted their locus of operations within the region. Loggers in Philippines had largely depleted the
nation's primary forests by the late 1990s, so their industry went into
decline (Dauvergne, 1998:121). Having harvested most of the accessible stands of timber in peninsular Malaysia, Malaysian logging firms
looked abroad for new sources of raw material, expanding their operations in Papua New Guinea and the Solomon Islands (Dauvergne,
1998:123). When the government imposed a ban on logging in Thailand, Thai logging firms moved their operations across the border
into Myanmar. In Sabah, Sarawak, and the outer islands of Indonesia,
the logging of lowland rain forests continued with unabated intensity
throughout the 1990s (Jepson et al., 2001).

The meta-analysis in part B of table 9.1 summarizes the results from
21 case studies of tropical deforestation in Southeast Asia during the
1990s. Like the studies in part A, these studies represent a wide array of
rural places in Southeast Asia. Vestiges of the earlier pattern of deforestation remained as indicated by the third causal conjuncture in part B,
in which logging and the associated road building opened up regions for
spontaneous colonists who then cleared the land. The other causal conjunctures indicate a fundamental shift in the forces driving deforestation
during the 1990s. Combinations of logging and plantation expansion
became more salient from the 1980s to the 1990s. This shift signals the

growing importance of large commercial enterprises in the coalitions of actors driving deforestation in Southeast Asia.

Entrepreneurs in the Southeast Asian logging industry, first in Philippines and later in Indonesia and Malaysia, have relied on the well-known principles of "crony capitalism" to organize the extraction of valuable wood from Southeast Asia's forests. Wealthy entrepreneurs with political connections used their personal influence to acquire logging concessions. Frequently, they would subcontract the actual exploitation of the concessions to local businessmen, for whom they would serve as patrons in the larger political and economic arena (Dauvergne, 1997). In Indonesia, these subcontracts weakened compliance because concession holders rarely required that subcontractors adhere to regulations about the volume of logs extracted from a concession (McCarthy, 2000) or the procedures for reforesting cutover areas. The same arrangements also allowed concession holders to blame subcontractors for violations in the logging of their concessions.

After the economic crisis and subsequent political turmoil in Indonesia during 1997, governmental control over logging activities declined even further. In the late 1990s, the central government devolved control over forest exploitation to the provinces. Devolution did not so much change the pattern of exploitation as it changed the participants in the patron–client networks that exploited the forests. In the late 1990s, local entrepreneurs in Sumatra paid local politicians and police to look the other way while gangs of loggers exploited forest concessions for which they had no licenses (McCarthy, 2000). In 1997, the volume of logs taken illegally from Indonesia's forests exceeded the volume taken legally from the forests (Sunderlin, 1999). The political–economic turmoil of 1997 temporarily increased the prospects of meaningful reform in logging practices and disrupted the clientilist logging networks, but the rapacious patterns of exploitation resumed shortly thereafter (McCarthy, 2000).

In the 1980s and 1990s, the intensified logging activities gave rise to other forest-destroying phenomena. The extensive amounts of slash (dead wood) left in postextraction secondary forests fed forest fires of unprecedented size and magnitude on Kalimantan and Sumatra during El Niño drought years in the early 1980s, late 1980s, and late 1990s (Dennis et al., 2001). Some large enterprises, often controlled by Malaysians, began to regard logging as only a stage in the exploitation of a tract of land. Company workers would log an area; then despite federal regulations against fires, workers would burn the remaining forest and plant oil palm on the burned-over tracts of land. These fires, especially those on "peat lands," had devastating environmental consequences (Tacconi, 2003). During the drought years of late 1990s, company-ignited fires burned large areas outside of the companies' concessions and created severe air pollution problems in Kuala Lumpur, Singapore, and other cities in Southeast Asia. They were "development fires" rather than wildfires (Stolle and Tomich, 1999).

Oil palm plantations increased tremendously in size in the outer islands of Indonesia after 1970. Between 1990 and 1999, the amount of acreage devoted to oil palm cultivation almost tripled in size. With almost 3 million hectares of Indonesia devoted to oil palm cultivation in 1999, Indonesia had become the second leading exporter of oil palm (after Malaysia) in the world (Casson, 2000). A high price, coupled with ample supplies of land and low labor costs, fueled the increase in oil palm acreage. International conglomerates figured centrally in the expansion of oil palm plantations in Indonesia. The state provided crucial assistance to the conglomerates by invalidating smallholders' claims to the same land that the conglomerates wanted to occupy. Although Indonesian law stipulated that the conglomerates should form "partnerships" with smallholders, few of them did so. Finnantara-Intiga, a Finnish–Indonesian joint venture, tried to work with smallholders in East Kalimantan, but the economic viability of their enterprise remained in doubt. Observing Finnantara's struggles, the heads of other oil palm enterprises ignored the regulations and relied on occasional state sponsored acts of coercion to secure their access to land and labor (Potter and Lee, 1998).

A comparison of the findings in parts A and B of table 9.1 about the forces driving deforestation before and after 1990 suggests that corporations may have eclipsed smallholders as the primary force driving deforestation in recent years. This conclusion is probably overdrawn, in that corporations were certainly vital contributors to deforestation processes during earlier periods as well. During the later period, corporations, by first logging an area and then planting it with oil palm, converted land from forests to fields without the assistance of smallholders (see PLANT in panel B of table 9.1). Although many sons and daughters of smallholders in Philippines, Indonesia, and Malaysia migrated to cities in the last decades of the twentieth century, others remained on the land. During the economic crisis of 1997, many of these smallholders expanded the area under cultivation at the expense of the forest in an attempt to generate extra income (Sunderlin, 1999). Nevertheless, the salience of

Table 9.2
Afforestation in Southeast Asia: A Qualitative Comparative Analysis

The 1980s and 1990s (51/54)
logging INTENSE SMSEDAG HILLS (4) +
logging INTENSE HILLS ROADS POP (1) +
INTENSE SMSEDAG HILLS roads pop (2) +
logging INTENSE SMSEDAG roads pop (1) +
logging intense smsedag hills roads POP (1)

The "+" in this table indicates that afforestation occurs whenever any of the listed combinations of causal conditions exists.

Upper case indicates that the causal factor is present, and lower case indicates its absence. The ratio in parentheses indicates the proportion of studies whose findings agree with the Boolean expressions in the lines below it. The number in parentheses after each causal combination indicates the number of studies whose findings fit that causal conjuncture.

Factors: HILLS, marginal terrain, ill suited for agriculture; INTENSE, agricultural intensification occurring; LOGGING, logging; POP, rural population increase; ROADS, recent road construction in the region; SMSEDAG, small-scale, sedentary agriculture.

smallholder-driven deforestation declined in most places as larger, corporate entities assumed a larger role in destroying the forests.

In some places, small-scale, sedentary farmers became a force for restoring forests during the 1990s. Table 9.2 reports a qualitative comparative analysis of studies in Southeast Asia during the 1980s and 1990s that report afforestation. Smallholder agriculture provided a common element in all of the cases, and marginal terrain figured in all but one case of afforestation. In Vietnam, smallholders intensified rice cultivation on permanent fields in the valleys at the same time they allowed afforestation to occur in the hills above the valleys (Sikor and Truong, 2002; Müller and Zeller, 2002). People in poor, mangrove-depleted coastal communities in central Philippines have begun planting mangroves in newly formed tidal mud flats formed by alluvial deposits close to the mouths of rivers (Walters, 2003). Extensive areas of regenerating "bush" emerged in the upland portions of Cebu in Philippines (Kummer, Concepcion, and Cenizares, 1994). Government subsidies and "regreening" campaigns facilitated afforestation in a significant number of these instances (Nibbering, 1999).

Smallholders have also begun to afforest *Imperata cylindrica* grasslands. During the 1970s and 1980s, population growth and continuing commercial opportunities persuaded smallholders to reduce the length of fallow periods throughout the region. Under these circumstances, *I. cylindrica* grasses spread and prevented secondary forests from reemerging in fallowed fields. By the 1990s, *I. cylindrica* grasslands had replaced for-

---

Marco Flores: From Cutting to Propagating Native Tree Species in the Philippine Uplands

An indigenous person with only a second-grade education, Marco Flores has spent much of his life working as an itinerant laborer in upland regions on the island of Negros. He has panned for gold, cut sugar cane, and for many years climbed all over the mountains of Negros looking for commercially valuable trees that logging companies might cut. During the early 1960s, he became friendly with an elder from his tribal group, the Tumandok, and, before the elder passed away, he talked extensively with Marco about the native flora of Negros. In the 1980s, with the old growth forests reduced to degraded fragments, the logging firms moved away from Negros, and Marcos joined fledgling local efforts to preserve native flora. He became the caretaker for a 9.5-hectare communal forest. Drawing on his extensive knowledge of the local biota, he has since stocked the forest with native tree species as well as mahogany and gmelina trees. He sells the native tree seedlings to local farmers and has recently begun to make furniture from these trees.

In presentations throughout Negros, Marco has argued for the superior qualities of native tree species in comparison with exotic species. He has also urged upland peasants to make furniture rather than charcoal from the wood in secondary forests, arguing that furniture construction creates more value for the impoverished people of the upland communities. Marco's transition from work as a scout identifying trees for felling to work as a conservationist intent on restoring native tree species reflects the gradual shift from a condition of forest abundance to forest scarcity that has occurred throughout much of Southeast Asia.

SOURCE: Garcia and Mulkins (1997).

---

ests on more than 40 million hectares in Southeast Asia (Hairiah, 2000), with particularly extensive grasslands in the Lesser Sunda Islands, Central Sulawesi, and the drier regions of Myanmar and Thailand. During the early 1990s, smallholders began reclaiming *I. cylindrica* grasslands in Sulawesi for agroforestry by using herbicides to suppress weed growth until the canopy created by cacao plants closed after several years of growth and shaded out the grass (Ruf, 2001:308–310).

In other locales, smallholders intensified agriculture on prime agricultural land in the lowlands and, in so doing, reduced shifting cultivation in adjacent upland locations. Smallholders in the coastal lowlands of Palawan in Philippines installed irrigation systems in the early 1990s, began to grow additional crops of rice on their irrigated lands, and hired more labor from off the farm to cultivate the crops. The newly hired labor-

Figure 9.2   Afforestation with eucalyptus trees in the hills above Quang Tri, Vietnam. **Photo credit: FAO photo (1994).**

ers had practiced shifting cultivation in the nearby uplands. With more paid labor now available in the lowlands, the farm workers reduced the amount of forest that they cleared each year in the uplands (Shively and Martinez, 2001). Marco Flores' life story (see box) testifies to the growing interest in afforestation in another upland area of Philippines.

In northern Vietnam, changes in land tenure during the late 1980s sparked a similar set of changes in the landscape. In a series of directives, the government privatized agriculture, giving peasants long-term leases for particular tracts of land. The government also expanded agricultural markets by allowing interprovincial trade of basic commodities such as rice. After the reforms, smallholders intensified rice cultivation in the lowlands and reduced the number of fields in agriculturally marginal upland areas. Smallholders allowed some of these lands to regenerate into natural forest; on other tracts, they planted fruit and timber trees as well as perennials such as coffee and tea (Tachibana, Nguyen, and Otsuka, 2001). The lands depicted in Figure 9.2 typify the areas currently undergoing afforestation in Vietnam.

With the shift in the forces driving deforestation in Southeast Asia after 1990, analysts have begun to reassess the role that smallholders play in forest cover change. To quote one prominent analyst (Fox, 2001), "Shifting cultivation, rather than being the hobgoblin of tropical conservation, may be ecologically appropriate, culturally suitable, and under certain circumstances the best means available for preserving biodiversity in the region."

Shifting cultivation looked good in this analyst's eyes because the alternative in Southeast Asia in the 1990s appeared to be corporate-driven conversion of lowland forests into plantations after logging. Shifting cultivation can, however, degrade landscapes. In one study of landscape change in northern Vietnam, entitled "Shifting Cultivation Without Deforestation," open-and-closed-canopy secondary forests declined from 65 to 18 percent of the land area between 1952 and 1995, whereas mixed grasses, bamboo, and woody shrubs increased to from 27 to 67 percent of the landscape during the same period (Fox et al., 2001). In other words, the sustained pressure of shifting cultivation on secondary vegetation can cause it over time to degrade from secondary forest into a fire-adapted vegetation dominated by shrubs and grasses. In an African context, this type of vegetation is called farm-bush. The large areas of "disturbed forest" in the upland interior of Southeast Asia in Figure 9.1 indicate how extensive this plant formation has become.

Even in this disturbed form, the landscapes associated with shifting cultivation probably contain more biodiversity than the lowland plantations established by corporations. Ironically, this reassessment of the ecological effects of shifting cultivation has occurred at a time when the numbers of shifting cultivators appear to have declined significantly (Padoch and Coffey, 2003). In some places, they have sedentarized; in other places, they have been evicted from lowlands by corporations or pulled out of the uplands by more attractive economic opportunities elsewhere.

Accentuated topography has played a role in the preservation of the most extensive insular block of forest remaining in the region, on the island of New Guinea. The most accurate estimate of forest-cover change in Papua New Guinea puts it at –0.2 percent per year between 1976 and 1996 (Wunder, 2003). Population growth among subsistence cultivators in the interior uplands has accounted for most of the deforestation during this period. The failure of the government to expand or maintain roads after 1970 has retarded deforestation rates in the interior highlands by making it costly to transport products to markets in the cities and overseas (Wunder, 2003). The government has spent little money on rural infrastructure, in part because it does not rely on rural road users for revenue. The state earns most of its revenue from mining enterprises, so it feels little need to foment agricultural development in the interior or to tax people living in the interior.

Logging in Papua New Guinea increased during the 1990s to almost 3 million cubic meters annually, as loggers depleted primary forests elsewhere in Southeast Asia and found a fresh supply of old growth forests in New Guinea. Selective logging in New Guinea usually did not initiate the now familiar forest-destroying sequence of logging followed by farming. Between 1975 and 1996, people converted only about 15 percent of the logged-over lands to nonforest uses (Wunder, 2003). The relative absence of "follow-on" farmers probably stems from the geographic separation of most smallholders from the logging activities. Logging occurs in coastal zones, whereas smallholders reside in the uplands. As a result, most logged-over stands of forest regenerated, albeit in a degraded form.

Conclusion: Trajectories of Change on Arable and Marginal Lands

Two distinct trajectories of land-cover change, one for marginal lands and another for arable lowlands, emerge from this analysis. In the lowlands, the extraction of logs from primary forests and the conversion of these lands to plantations shows no signs of slowing down. The economic crisis of 1997 may have slowed the conversion of these lands, but it resumed shortly thereafter on a large scale (Casson, 2000). The continued growth in internal and in some cases external markets for tropical hardwoods and oil palm, coupled with the continued weakness of the state, makes it hard to conceive of how old growth forests might continue to survive in the lowlands of Southeast Asia (Jepson et al., 2001).

The rugged topography that covers much of Southeast Asia provides a more hopeful note for conservationists, because a greener trajectory of landscape change may be emerging in the region's marginal lands, usually in upland settings. The limited agricultural productivity of inaccessible and steeper-sloped lands reduces the marginal utility of farming on these lands. Smallholders seem likely to reassess the viability of upland agriculture if the rapid economic growth and urbanization of the past two decades continues in Southeast Asia. Under these circumstances, upland farmers may choose to work in intensified lowland agricultural enterprises or expanding urban labor markets. If they opt for either of these choices, they will labor less on upland farms, most likely by reducing lands devoted to more labor-intensive agricultural uses and increasing lands devoted to less labor-intensive forests and agroforests.

Afforestation in the uplands and deforestation in the lowlands corresponds roughly to the expectations of forest transition theory (Mather and Needle, 1998), and points implicitly to the importance of trying to manage the two contrasting patterns of landscape change in ways that favor biodiversity conservation. The plans for biodiversity conservation in Southeast Asia are well conceived, but implementation has usually been politically difficult (Mittermeier et al., 2000:290). Under these circumstances, reforms that strengthen the state would enable it to implement these plans more effectively. In the lowlands, low-impact logging techniques would improve the health of postextraction secondary forests. Genetically improved rubber trees would increase the prosperity of thousands of smallholders in the lowlands, create more "jungle rubber" (de Jong, 2001), and provide smallholders with greater incentives to resist companies that try to appropriate their lands.[5] In the uplands, government-assisted afforestation efforts, especially in the buffer zones of parks, should facilitate the expansion of forest cover in these settings.

Through a Regional Lens: Conservation Policies in Large and Small Forests

The brief twilight commenced, and the sounds of multifarious life came from the vegetation around. The whirring of cicadas; the shrill stridulation of a vast number and variety of field crickets and grasshoppers, each species sounding its peculiar note; the plaintive hooting of tree frogs—all blended together in one continuous ringing sound—the audible expression of the teeming profusion of nature. As night came on, many species of frogs and toads in the marshy places joined in the chorus: their croaking and drumming, far louder than anything I had heard before in the same line, being added to the other noises created an almost deafening din. This uproar of life, I afterwards found, never wholly ceased, night or day: in the course of time I became, like other residents, accustomed to it.
HENRY WALTER BATES, *The Naturalist on the River Amazons* (1969 [1864])

The din arises not from an undisturbed jungle but from insects and frogs living amid the plants in a nineteenth-century urban neighborhood in Belem, a city at the mouth of the Amazon River. Bates's observation suggests that impressive amounts of biodiversity can live in close proximity to large numbers of humans. It makes more plausible the essential contention of sustainable development efforts—that humans can maintain the integrity of tropical nature at the same time that they exploit it for their own purposes. The salience of the frogs, currently one of the most endangered genera, adds to the poignancy of the passage and underscores how difficult it may be to achieve such a laudable goal.

Policies pursued by nongovernmental organizations, corporations, and governments represent the tools of sustainable development. The different regional deforestation processes and enabling conditions documented in previous chapters suggest that the optimal mix of policies for advancing biodiversity conservation will vary substantially across the seven rain forest regions. To promote biodiversity conservation effectively, policies must recognize these variable human ecological conditions and build on them. Unfortunately, academic discussions about the relative merits of different policies quite frequently fail to clarify these contextual factors. Arguments often proceed in a highly abstracted way. Writers make a case for or against a particular policy in their disciplinary vernacular and sometimes illustrate their argument through the examination of a case that supports the argument. Although this mode of argument has undoubted advantages in clarifying sequences of cause and effect, it leaves ambiguous the range of places to which the argument applies. For example, analysts have established that loggers initiated processes of tropical deforestation in Southeast Asia during the 1970s and 1980s. No one went on to explain why the same process occurred in West Africa, but not in central Africa, Central America, or Amazonia. In other words, many arguments about changes in tropical forest cover and related policies are not bounded geographically. This shortcoming makes it difficult to assess the potential of a particular policy innovation for reducing rates of deforestation in a particular place. The last few pages of the book use the preceding regional analyses to put geographic boundaries around the drivers of forest decline and the policies that might slow the decline. A brief review of the different regional processes of forest-cover change and a comparison of these processes with those described in the Geist and Lambin metaanalysis (2001, 2002) should set the stage for a discussion of how the geographic boundaries of effective policies for biodiversity conservation might vary.

The Regional Patterns: A Summary

Some policy-relevant processes of forest-cover change characterize all regions. Markets consistently undervalue the environmental services that primary forests provide, thereby creating an economic bias against conservation (Balmford et al., 2002). Norms encouraging the excessive consumption of tropical commodities find their fullest expression in the affluent Western nations, but they exert pressure on forests throughout the world. Rural-to-urban migration has begun to deprive rural enterprises throughout the developing world of workers to convert forests into fields, and at the same time the number of new urban consumers who purchase products from these fields has increased. Finally, the growing importance of urban political agendas, coupled with critiques from environmentalists and International Monetary Fund (IMF)-induced fiscal pressures, has persuaded governments to abandon colonization projects throughout the tropical world.

Other factors with global reach vary dramatically from region to region and create unique constellations of forces that drive forest-cover change in each region. Globalization has accentuated regional differences. The geographic expansion of free trade in crops, minerals, and timber has gradually dismantled the passive protections against exploitation that have existed in central Africa and South America. For example, during the 1980s, no one logged forests in the Guianas in northeast South America. Then the market expanded to include the Guianas, and companies from the other side of the globe began to extract timber from these places. By opening these regions up to new producers, globalization puts new competitive pressures on producers in old agricultural zones and may induce them to abandon old fields, mines, or forest concessions. In this way, globalization accelerates the destruction of old growth forests in some regions at the same time that it expedites the emergence of secondary growth in other regions. In general, globalization increases human pressures on the forest-rich regions such as the Amazon basin and reduces human pressures on forest-depleted regions such as West Africa or south Asia (Rudel, 2002).

Both the Geist and Lambin (2001) study and this analysis identify distinctive regional configurations of causes: cattle-raising complexes in Latin America, logging-initiated processes of deforestation in Southeast Asia, and smallholder agriculture in Africa. The patterns of change identified in this study, their driving forces, and the appropriate conservation policies are outlined in table 10.1. Both the Geist and Lambin study and this study conclude that population increase has been overemphasized as a causal force. In this analysis, it drives deforestation in five regions during the 1980s (West Africa, central Africa, East Africa, south Asia, and Southeast Asia) but in only one region during the 1990s (East Africa).

This last observation points out two important differences between this study and the Geist and Lambin analysis. First, this study looks at historical trends in the causal configurations, whereas the Geist and Lambin study does not. Probably the most important historical finding concerns the decline in the role of the central state between the 1980s and the 1990s in processes of forest-cover change. During the 1970s and the 1980s, governments on all three continents sponsored new land settlement schemes and constructed penetration roads to open up the newly settled areas for commerce and agriculture. By the end of the 1990s, only Brazil maintained a new land settlement program, and it operated on a much reduced scale compared with the 1970s. In the meantime, corporate enterprises had become more salient in deforestation processes on all three continents. Logging firms and soybean producers in the Amazon basin, timber companies in the central African forests, and logging–plantation companies in insular Southeast Asia all became more salient in forest destruction. Over the course of the past two decades, state-assisted deforestation has evolved into enterprise-driven deforestation in tropical regions with large forests.

Second, the use of the qualitative comparative analysis (QCA) method in this study makes it more sensitive to subregional variations in causal configurations than studies, such as the Geist and Lambin study (2001), that group cases by continent. In earlier work using QCA (Rudel and Roper, 1996), we found seven distinct regional configurations. Forests vary dramatically in size across the seven regions (table 10.2). Three regions have large forests (South America, central Africa, and Southeast Asia) and four regions have comparatively small forests (Central America, West Africa, East Africa, and south Asia). With several significant exceptions (Colfer and Byron, 2001), analysts have ignored the impact of forest size on the effectiveness of conservation policies. The causal configurations featured in this analysis underscore the importance of forest size in shaping regional patterns of forest-cover change. The Latin American analysis splits into regions with large (South America) and small (Central

Table 10.1

Regional Patterns of Forest-Cover Change and Their Policy Implications

Regions	Forest-Cover Change	Driving Forces	Conservation Policies
Central America and the Caribbean	Small forests: slowing but still rapid decline; slowing attributable to afforestation	Growing urban markets drive forest decline; emigration and tourism promote afforestation.	Expanding farm forests; certification; agricultural intensification; ICDPs
South America	Large forests: moderate rates of deforestation, but amounts of forest lost are large	Passive protection erodes; logging, agriculture, and fires expand; resource partitioning occurs.	International forest service payments; reserves; easements; enterprises engaged in ecological modernization
West Africa	Small forests: continued rapid deforestation	Growing urban markets; continuing agricultural exports; debt	Expanding farm forests; agricultural intensification; ICDPs
Central Africa	Large forests: little deforestation; peri-urban forest losses and peripheral forest gains	Passive protection: inaccessibility; mining economy suppresses farming	International forest service payments; certified logging; ICDPs
East Africa	Dry forests with little biomass: net losses; peri-urban forest gains and peripheral forest losses	Charcoal production; domestic market expansion; rural population increase	Expanding community and farm forests; agricultural intensification; ICDPs
South Asia	Small forests: net afforestation	Charcoal production; control over forests ceded in part to villages	Expanding village and farm forests; agricultural intensification; ICDPs
Southeast Asia	Large forests: rapid decline in lowlands; some recovery in highlands	Large-scale logging, agriculture and fires without regulation	International forest service payments; agricultural intensification; ecological modernization

ICDP, integrated conservation and development programs.

Table 10.2
Sizes of Tropical Forests by Region in 2000

Region	Forest per Capita*	Percentage Forested†
Central America and the Caribbean	0.75	30.82
South America	12.02	62.38
West Africa	0.88	31.32
Central Africa	7.05	58.83
East Africa	1.00	26.56
South Asia	0.10	22.27
Southeast Asia	1.79	54.10

*Values (in hectares) are means for the countries in the region.
†Values are the regional means. In evaluating the significance of the differences between regions, note that the Food and Agricultural Organization uses a generous definition of *forest:* any place where the canopy exceeds 10 percent.

America) forests. The African analysis also divides into two regions with sparse or small forests (East and West Africa) and one region with large forests (central Africa). Similarly, Asia divides into south Asia with small forests and Southeast Asia with large forests. These geographic divisions throw into bold relief the differences between large and small forests in processes of forest-cover change. For example, whereas forest-rich central Africa exhibits peri-urban deforestation and remote rural reforestation, forest-poor East Africa experiences peri-urban afforestation and remote rural deforestation. The next section outlines the implications of these differences in forest size for biodiversity conservation in the tropics.

The Size of Forests: Policy Consequences

As tropical forests have declined in size during the past half century, agents of forest-cover change have altered their behaviors in ways that have important implications for conservation policy. Table 10.3 summarizes these effects, and the following pages describe each of them in more detail.

Road Building and Other Transportation Improvements

Large tropical forests exist in places beyond the reach of the commercial world. Although the extent of this forested domain continues to decline as transportation improvements bring these places closer to urban consumers, portions of western Amazonia, the central Congo basin, and

Table 10.3

Forest-Cover Dynamics in Large and Small Forests

Driving Forces and Conservation Policies	Effects on Large Forests	Effects on Small Forests
Road building and other transportation improvements	Breakdown of passive protections; improvement in access to markets; deforestation	Improvement in access to markets; afforestation encouraged
Corporate logging and agricultural expansion	Destruction or degradation of large tracts of forests	Effects minimal, because not enough commercial value to logging or expanding agriculture in this setting
Conservation strategy: farm and community forestry	Usually not viable	Viable, especially close to villages
Conservation strategy: parks, easements, and integrated conservation and development projects	Politically viable and justifiable only in this setting	Too coercive to be political acceptable or justifiable in this setting
Forest transitions	Occur only when nonfarm economic opportunities grow rapidly	Occur when the growing scarcity of wood products encourages afforestation

the swamp forests of eastern Indonesia remain so remote from urban markets or so difficult to develop for commercial use that they remain relatively unexploited despite the absence of legal protections. States and environmental nongovernmental organizations (NGOs) do nothing to protect these forests, but they continue to exist because physical conditions prevent their exploitation. In this sense, these forests experience a kind of passive protection.

Regions with small forests such as the densely settled rural areas in south Asia, East Africa, and West Africa also contain forests far from urban centers in places without roads. The remoteness of these forests does not protect them. Roads do not serve these areas, but the presence of smallholder settlements within walking distance of the forests, coupled with local scarcities of land and wood products, makes the forests vulnerable to degradation and deforestation. People will walk long distances to harvest natural resources in these places, and the institutional protections for forests provided by villages and the state frequently do not reach that far, so conditions of open access prevail and people degrade the resource.

Transportation improvements change these conditions in both forest-rich and forest-poor regions. When contractors build roads through forested land in forest-rich regions, large and small landholders immediately try to capitalize on the improved access to urban markets. They log the nearby forests and clear the land for agriculture, creating corridors of cleared land along the new roads. When governments pave dirt roads, access improves again, and another round of forest clearing begins. In places with small forests and extensive tracts of agricultural land, road improvements touch off a different dynamic. The easier access to urban markets makes roadside land more expensive, which in turn encourages owners to intensify their agricultural operations on these lands. In Vietnam, road improvements in mountainous regions spurred the intensification of wet rice agriculture along the roads in the valleys. The additional labor demands of the intensified rice cultivation persuaded smallholders to abandon agriculture along the upper slopes of the valleys and create less-labor-demanding tree plantations on these lands (Sikor and Truong, 2001; Müller and Zeller, 2002).[1] In analyses of forest-cover decline in the Amazon basin between 1985 and 1995, road paving caused rapid land clearing when the newly paved roads traversed forested regions, but road paving usually occurred in already cleared areas, and in these places it caused agricultural intensification. In theory, intensified production from roadside lands should have reduced the incentives to clear forests from lands far from roads (Andersen et al., 2002:126–127).

Corporate Logging and Agricultural Expansion

Corporations and well-capitalized family enterprises drive the destruction of most rain forests in two of the three regions with large forests (South America and Southeast Asia), and in the third region with large forests (central Africa) they degrade forests through logging operations. During earlier historical periods, large enterprises either cleared or logged extensive tracts of rain forest in Central America and West Africa.

The parable of the little man and the big stone accurately depicts the situation of rain forests relative to the well-financed organizations that would exploit them. In the parable, poor forest dwellers who find a big stone (gem) rather than a small stone eventually suffer because their good fortune attracts the interest of more powerful people who find a way to take the stone from the forest dweller. Forest dwellers with small stones never attract the attention of the "big man," so they do not lose their stones (Dove, 1993:17–18).

So it is with forests. Extensive forests contain large repositories of wealth in the form of standing timber and unexploited soils, so they attract the interest of large corporate investors and wealthy families, intent on maximizing profits (Schnaiberg and Gould, 1994), who proceed to devastate the forests while extracting wealth from them. Smaller or more fragmented forests in places such as Central America or south Asia no longer contain large volumes of valuable natural resources, having lost them during earlier periods of exploitation. Corporate investors and wealthy families show little interest in exploiting these forests.

The crucial role of corporations in the exploitation of large forests probably explains why deforestation rates in the Amazon basin and insular Southeast Asia have risen and fallen in unison during the past two decades (Skole, 1999). Booming economies and easy credit spur these firms and their associates to destroy more forest. Efforts to reduce the collateral damage to the forests caused by these operations, frequently through the introduction of new technologies (ecological modernization), have an impact only on the large forests where the large corporations work. When the amount of commercially valuable timber in old growth forests falls below a certain threshold, logging firms abandon these forests. The declining salience of logging firms as drivers of deforestation in south Asia may be attributable to loggers' decisions to shut down their operations in the small, remnant tracts of old growth forests. In the aftermath of the large firms' departure, small teams of locally based loggers may mine the forests for additional wood or begin tree plantations (Pinedo-Vasquez et al., 2001).

Community Forestry, Farm Forestry, and Environmental Service Payments

Where forest resources are abundant and people are poor, arguments for forest conservation fall, for good reasons, on deaf ears. Residents of forest-rich regions see the forest as a resource to be exploited in the quest for higher standards of living. Given this orientation, both community- and state-run forest conservation programs encounter resistance in forest-rich regions. There are several exceptions to this pattern. If indigenous peoples derive their subsistence from the forests, they too may work to preserve large tracks of standing forest, as some groups of Shuar have done in Ecuador (Rudel with Horowitz, 1993). Other residents of forest-rich regions support the idea of income-producing forests, such as the proposed network of forest reserves in the Brazilian Amazon (Verissimo, Cochrane, and Souza, 2002), but these schemes protect only a portion of the forests. International environmental service payments organized around agreements such as the Clean Development Mechanism of the Kyoto Protocol offer an additional way to preserve large blocks of forest (Pfaff et al., 2000; Schultze, Wirth, and Heimann, 2000).

As forests continue to decline in extent, local residents become more aware of the environmental and economic costs of forest decline, and they show a greater willingness to work to reverse these trends (Uphoff, 2001:441). During the past 15 years, rural inhabitants of forest-poor communities in India have shown much more interest in regenerating cutover forests through joint forest management than have the inhabitants of forest-rich communities on the subcontinent (Poffenberger, McGean, and Khare, 1996:38). Smallholders usually do not intend to preserve or restore biodiversity in an area when they create farm or community forests, but, by conserving forest, soil, and water resources, they create the conditions for conserving biodiversity in an area (Uphoff, 2001:449). The success of farm and community forest projects often depends on other, related changes in a region. Improved and devolved governance, environmental service payments, integrated conservation and development programs (ICDPs), and increases in nonfarm economic opportunities all increase the likelihood that smallholders will conserve, rather than exploit, local forests.

During the 1990s, governments on all three continents began assigning regulatory powers over forests to local governments (Kaimowitz, Vallejos, et al., 1999). Insofar as these forest reforms allocate real regulatory power to local communities, they establish more participatory political processes. The effects of these reforms on biodiversity conservation could well vary with the size and value of the forests to be regulated. In Indonesia, wealthy investors have had a keen interest in exploiting the remaining lowland dipterocarp forests. When a forest reform endowed local governments with regulatory powers over forests in 1996, investors included local officials in the web of corrupt transactions that enabled them to exploit the forests with impunity (McCarthy, 2000). In contrast, large commercial land and logging firms have no interest in the cutover and degraded forest districts in India, so devolution in these places promises to create a more participatory, hands-on regulation of re-generating forests by the people who live around the forest (Poffenberger and McLean, 1996). Under these circumstances, devolution should, if anything, protect regenerating forests. In sum, devolution promises to ac-centuate the divergent trends in forest growth between places with large forests and places with small forests.

Parks, Easements, and Integrated Conservation and Development Projects

During the late 1980s, in response to widespread concern about bio-diversity conservation, NGOs and development assistance agencies launched integrated conservation and development assistance programs that promised to conserve biodiversity in particular locales by recruiting local residents into occupations that either sustained or at least did not destroy forests. Since 1995, conservationists have gradually become dis-enchanted with ICDPs, arguing that they usually do not protect plants and animals in tropical parks and their buffer zones (Terborgh, 1999; Brandon, 1998; Barrett and Arcese, 1995). Increasing numbers of conser-vationists have begun to argue that governments and foundations should purchase tracts of forest and hire guards to protect the new parks and reserves (Pimm et al., 2001). Related research shows that increases in the number of park guards improve the degree of environmental protection provided by parks (Bruner et al., 2001). The purchase of easements to protect forests from logging, or the outright purchase of threatened for-ests, has begun to seem like a more feasible conservation strategy since research by Conservation International established that "25 global bio-diversity hotspots," covering 1.4 percent of the Earth's land area, contain 44 percent of all vascular plant species. These geographic concentrations of biodiversity allow the conservation movement to focus their financial and human resources on conserving a few forests instead of an entire biome (Mittemeier et al., 2000).

   In some instances, the purchase of conservation easements—in effect, conservation concessions—has included provisions for creating jobs for indigenous people living in the concessions (Abelson, 2000), but in most instances the new direction in conservation policy provides little in the way of benefits for the people who live in or around tropical forests. Steve Brechin and his colleagues have criticized the new direction in con-

servation policymaking because, by preventing the poor from using forest resources, it violates elementary principles of social justice (Brechin
et al., 2002; Wilshusen et al., 2002). Common-property theorists make a
similar point (Poteete and Ostrom, 2002).

Clearly, the new conservation policies make sense only in remote, already depopulated forest zones. The frequent claim that there are no
"empty places" in tropical forests is certainly true, but this argument
should not be allowed to obscure the large differences in population densities between, for example, a rapidly growing frontier region such as
Rondônia, Brazil, and more remote regions in the northwestern Amazon
basin that are beyond the margins of commercial cultivation. In remote
rural places, conservation concessions with provisions allowing indigenous peoples to pursue traditional livelihoods represent an appropriate land use. Adaptive co-management strategies that bring indigenous
in-holders together with concession holders might provide the appropriate institutional framework for managing these concessions from year to
year (Ruitenbeek and Cartier, 2001). Conservationists might complain
that this strategy does not preserve the most threatened forests. This ob-
jection is accurate, but, as infrastructure projects such as Avanca Brazil
bring forests in remote places closer to markets, concessions should begin
to prevent outsiders from exploiting forests inside their borders. In effect,
conservation concessions and parks in remote rural places represent a
"cold spot" strategy in that they focus conservation efforts on forest-rich
regions that are not experiencing much development pressure.[2]

More settled places require different conservation strategies, quite
frequently involving some type of common property. In the past two
decades, analysts concerned with common property have developed a
theory and amassed evidence that local communities can manage forest
resources sustainably (Gibson, McKean, and Ostrom, 2000; Stone and
D'Andrea, 2001; Poteete and Ostrom, 2002). When resource users have
traditions of local governance, when they derive a considerable portion
of their income from the forest, and when the forest is small, bounded
and easily monitored, then collective action by local residents to manage
forests becomes more likely (Poteete and Ostrom, 2002).[3]

Typically, the common-property theorists present their arguments for
community-based natural resource management (CBNRM) in two steps.
They outline in the abstract why rational actors would want to participate in community-based natural resource management, and then they
present case studies that illustrate their argument. Because the facilitating

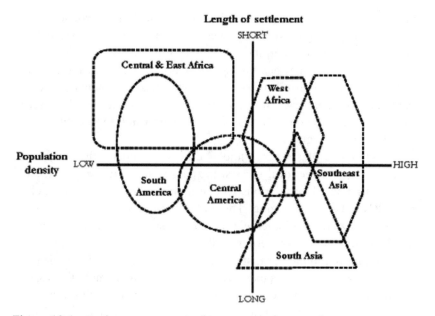

Figure 10.1   Settlement patterns in the seven rain forest regions.

conditions noted earlier occur unevenly across the different rain forest regions, CBNRM should be more prevalent in some regions than others, but common-property theorists never note these regional differentials. This omission gives their arguments more universal application, but it also endows them with the hegemonic overtones of microeconomic arguments. Recent conservationist arguments suffer from the same failing. Terborgh (1999) makes large claims based on several case studies. The resulting "purchase plus guards" formula seems to have universal application. The analysis here tries to transcend this "one size fits all" tendency in the literatures.

The institutions outlined here work best in different regional situations. As common-property theorists note (Poteete and Ostrom, 2002), CBNRM works best when communities have preexisting traditions of local governance. However, these traditions do not exist in many places. They are most likely to arise on long-settled, densely populated places beyond the reach of a central authority (Carneiro, 1970; Colfer and Byron, 2001). Many south Asian villagers developed these traditions through a long period of self-governance before the British established a colonial regime (Guha, 2000). As figure 10.1 demonstrates, most other rural regions in the tropics did not have long-established rural settlements, and therefore they did not experience these enabling conditions for local self-governance. Slave raiding and disruptive colonial resettlement schemes uprooted rural populations throughout tropical Africa during the nineteenth and twentieth centuries. In the 1960s, governments in Central America, South America, and Southeast Asia launched new land settlement schemes that created thousands of new rural communities composed of recent migrants.

Communities of strangers find it difficult to create the institutions necessary to manage local forests. In sparsely populated zones, the difficulties that small populations face in monitoring forests prevent effective CBNRM from emerging. These observations suggest that CBNRM promises to work best in densely settled, forest-depleted places with long traditions of some self-governance. People in rural south Asia and in the inner islands of Indonesia satisfy both of these conditions, and large numbers of these communities have recently begun to implement CBNRM.

Integrated conservation and development programs differ from CBNRM arrangements in that outside entities, most frequently aid agencies and environmental NGOs, provide subsidies for conservation strategies in and around endangered habitats. These projects begin from the economically plausible premise that local resource managers usually cannot withstand the economic pressures applied by well-financed economic interests who want to exploit endangered habitats. Under these circumstances, ICDPs represent the only hope that conservationists have for saving endangered habitats because, through their subsidies, ICDPs provide a countervailing economic influence that might persuade locals to conserve nearby forests. Even then, the subsidies may not be adequate. For example, in the 1990s, politically connected loggers enlisted village headmen in logging operations in and around the Gunung Leuser National Park despite the presence of a World Wildlife Fund–sponsored project that provided assistance to village councils in the conservation of their forests (McCarthy, 2000).

Variations across the different rain forest regions in the prevalence of largely forested, sparsely populated landscapes and densely settled, predominantly agricultural landscapes imply differences in policy emphases across the seven regions. CBNRM schemes would appear to have wide application in south Asia and East Africa with their long-exploited agricultural landscapes. Conservation easements and parks would appear to have more utility in central Africa and the Amazon basin than elsewhere, because these regions have large and remote expanses of forest. ICDPs may represent one of the few ways to prevent the total destruction of the small and endangered tropical habitats of West Africa and Central America. Each policy tool has its place in particularly diverse rain forest regions such as Southeast Asia. CBNRM schemes should help preserve or restore upland forests in continental Southeast Asia, and ICDPs may represent the only hope for conservation in the rapidly deforesting frontier regions of Kalimantan. Conservation easements may provide a useful way to preserve the forests of inland Irian Jaya without disrupting the lives of indigenous peoples.

Forest Transitions

During the past two centuries, a small stream of rural, agrarian societies have become urban, industrial societies, beginning first in Europe, extending early in the twentieth century to North America, and more recently to countries such as Brazil and Malaysia in the tropical realm. In many instances, a sequence of changes in forest cover has accompanied this "great transformation" in human livelihoods (Polanyi, 1944). Initially, population increase and agricultural expansion reduced the forest cover to low levels, but when young people began to leave rural areas for industrial jobs in cities, a turnaround in forest-cover trends occurred. Deforestation ceased and net afforestation began. Alexander Mather and others refer to this historical sequence of changes in forest cover as "the forest transition" (Mather and Needle, 1998). For other analysts, the changes in forest cover point to the existence of an "environmental Kuznets curve," in which sharp declines in forest cover give way to modest increases in forest cover as economic development occurs (Mather, Needle, and Fairbairn, 1999; Erhardt-Martinez, Crenshaw, and Jenkins, 2002). The abundant evidence for forest transitions in the now affluent, industrialized societies raises questions about how comparable processes might unfold as societies in the tropics undergo urbanization and industrialization. Would a full-fledged tropical forest transition reverse the current declines in biodiversity and render the policy discussions in this chapter superfluous?

The most recent data from the Food and Agricultural Organization (FAO) on forest-cover changes in tropical countries between 1990 and 2000 suggest that a forest transition has begun to emerge in a selective way in the tropics. Evidence for a transition is most abundant in south

Asia. India and Bangladesh experienced "turnarounds" in forest cover between the 1980s and the 1990s, going from net deforestation to net afforestation (FAO, 2001). The more affluent islands in the Caribbean also experienced unprecedented increases in forest cover during the same period (see chapter 3), and parts of Central America also began to afforest. Although these societies seem quite different in socioeconomic terms, none of them contains extensive stands of old growth forests. They all represent forest-depleted environments. The Amazon basin, the Congo basin, and insular Southeast Asia, the places with major blocks of rain forest, continued to lose forest cover in the 1990s much as they had during the 1980s, so they showed no signs of a forest transition. Because a tropical forest transition seems to be emerging only in already depleted environments, it does little to stem the worldwide losses of biodiversity in the tropics.

Although a forest transition cannot be expected to save large amounts of tropical biodiversity from extinction, it can restore degraded environments, sequester carbon, and prevent some extinctions. For these reasons, analysts might want to investigate how changes in agricultural and forest policies might accelerate the onset of the transition.[4] Combinations of changes may prove most effective. For example, agricultural intensification on prime agricultural lands in circumstances where all of these lands are cultivated should, through productivity increases and corresponding declines in the prices of crops, put pressure on farmers to convert marginally productive agricultural fields into forests (Angelsen and Kaimowitz, 2001). Farmers are more likely to make these changes if governments subsidize the purchase of seedlings from tree nurseries.

Conclusion

As self-conscious efforts to achieve sustainable development intensified during the 1980s and 1990s, conservationists and concerned scientists invented new ways to promote rain forest conservation in the tropics. People created conservation easements, integrated conservation and development plans, forest certification schemes, debt-for-nature swaps, joint forest management programs, and the Clean Development Mechanism of the Kyoto Protocol. The multiplication in the means for conserving rain forests gives policymakers a wider array of policies from which to choose and challenges them to come up with an appropriate mix of policies for each region. Even if policymakers apply the new tools for conservation appropriately across the different regions, they will not stop the destruction of rain forests in lowland Indonesia or in the south central Amazon basin. Taken together, the innovations may produce small, but still significant, increments in the conservation of rain forests across a diverse set of tropical locales.

The criteria for judging conservation efforts successful should also vary across regions. For the forest-depleted regions of south Asia, East Africa, and West Africa, success may mean preserving biodiversity in a few small areas and restoring tree cover in the form of plantations over much larger areas. In Central America, success might involve widespread spontaneous regeneration of secondary forests in areas experiencing out-migration. In central Africa and the Amazon, success would mean the preservation of much larger areas as either undisturbed forests or as minimally disturbed, postextraction secondary forests. In Southeast Asia, with its diverse processes of landscape change, success should come in different forms—plantations in the uplands of continental Southeast Asia and reserves in the uplands of insular Southeast Asia.

Regionally differentiated policies and criteria for success should make it possible for conservationists to focus on "the art of the possible" in the struggle to conserve the world's remaining tropical forests. Given the inability of the world's governments to agree on and implement a joint program of action to preserve tropical forests (World Wide Fund for Nature, 1999), regionally appropriate mixes of policies, financed in part from external sources and implemented locally, offer the best chance of achieving meaningful amounts of conservation in the tropical realm.

Appendix: Case Studies and Accompanying QCAs for Each Region

Interpretive Notes

In an effort to make the regional analyses of the assembled case studies more intelligible and transparent, this appendix lists the case studies for each region and the causal configuration for each case in the qualitative comparative analysis (QCA) for that region. By consulting the regional QCA, one can figure out the place of a particular study in the logical expressions that summarize the conditions that drive deforestation or afforestation in the region. However, these causal configurations represent the lowest common denominators for the region, so the codes for each case frequently omit important conditions particular to that case. In interpreting these data, it is important to remember that the minimized logical expression includes only case studies that exhibit the phenomenon (deforestation, for example) for which the expression is being minimized. Studies that exhibit afforestation will not, in this example, appear in the QCA. A study may occur during both the 1980s and the 1990s if the authors made separate measurements of land-cover change during each period. In these instances, I refer to the study as, for example, "Schelhas94" and "Schelhas96," with the "94" and "96" referring to the year of publication. Only one publication per project is included in the list. The names listed are the first authors on the relevant publication. Where one author has carried out multiple studies, I add the last two digits of the year of the different publications to distinguish them from one another (e.g., Fearnside93). To save space, I have listed only first authors next to the corresponding causal configuration in the regional QCAs. In some instances, I have identified coauthors where it was necessary to identify a study. For those studies that are not readily accessible in journals or books, I have given more complete references to make them easier to find.

Central America and the Caribbean

Case Studies

Arizpe et al. 1996. *Culture and Global Change: Social Perceptions of Deforestation in the Lacandona Rain Forest in Mexico.*
Berti Lungo. 1999. *Transformaciones Recientes en la Industria y la Politica Forestal Costarricense.*
Brothers. 1997. *Environmental Conservation* 24(3): 213–223.
Camargo. 1984. In *Colonizacion y Destruccion de Bosques en Panama.*
Carr. 2000. *Colonization, Land Use and Deforestation in the Sierra de Lacandona National Park, Peten, Guatemala.*
Carriere. 1991. In *Environment and Development in Latin America.*
Catanese. 1993. *Canadian Journal of Development Studies* 14(1): 59–72.
Chomitz and Gray. 1996. *World Bank Economic Review* 10: 487–512.
Collier et al. 1994. *Bioscience* 44(6): 398–407.
Cortes and Restrepo. 1998. *Deforestation and Forest Degradation in the Region of the Black Communities of the Colombian Pacific.* Paper presented to the Latin American Workshop on the Underlying Causes of Deforestation and Forest Degradation, Santiago, Chile. Available at www.wrm.org.uy.
DeWalt. 1983. *Bulletin of the Atomic Scientists* January.
Dirzo and Garcia. 1992. *Conservation Biology* 6(1): 84–90.
Eyre. 1987. *Ambio* 16(6): 338–343.
Faris. 1999. *Deforestation and Land Use on the Evolving Frontier: An Empirical Assessment.*
Fuentes Aguilar and Soto Mora. 1992. *Revista Geografica* 116: 67–84.
Geoghegan et al. 2001. *Forest Ecology and Management* 154: 353–370.
Godoy et al. 1997. *World Development* 25: 977–87.
Halhead. 1992. In *Development or Destruction: The Conversion of Tropical Forest to Pasture in Latin America.*
Harrison. 1991. *Interciencia* 16(2): 83–93.
Hecht. 2004. In *Liberation Ecologies...*, 2nd edition.
Heckadon Moreno. 1984. In *Colonizacion y Destruccion de Bosques en Panama.*
Hernandez. 1984. In *Colonizacion y Destruccion de Bosques en Panama.*
Humphries. 1998. *Economic Development and Cultural Change.* 47(1): 95–124.
Janzen. 1988. In *Biodiversity.*
Joly. 1984. In *Colonizacion y Destruccion de Bosques en Panama.*
Jones. 1992. *Policy Studies Journal* 20(4): 694–697.

Klooster. 2000. *Yearbook, Conference of Latin Americanist Geographers* 26: 47–59.

Larson. 2000. *Peasant Agroforesters: Fiction or Reality: Colonization and Forest Conversion on the Nicaraguan Frontier.* See reference list for complete citation.

Lewis and Coffey. 1985. *Ambio* 14: 158–160.

Lobato. 1980. *Revista Ciencia Forestal* 5(24): 45–54.

Ludeke et al. 1990. *Journal of Environmental Management* 31: 247.

McKay. 1984. In *Colonizacion y Destruccion de Bosques en Panama.*

Mendoza and Dirzo. 1999. *Biodiversity and Conservation* 8(12): 1621–1641.

Monaghan. 2000. *Peasants, the State, and Colonization of Haiti's Last Rainforest.* See reference list for complete citation.

Nations and Komer. 1982. *Cultural Survival Quarterly* 6: 8–12.

Nygren. 1995. *Forest and Conservation History* 39(1): 27–35.

O'Brien. 1998. *Sacrificing the Forest: Environmental and Social Struggles in Chiapas.*

Ochoa-Gaona and Gonzalez-Espinosa. 2000. *Applied Geography* 20(1): 17–42.

Parsons. 1976. *Revista de Biologia Tropical* 24(1): 121–138.

Peckenham. 1980. *Latin American Perspectives* 7(2–3): 169–177.

Pierce. 1992. In *Changing Tropical Forests: Historical Perspectives on Today's Challenges in Central and South America.*

Place. 1981. *Ecological and Social Consequences of Export Beef Production in Guanacaste Province, Costa Rica.*

Price. 1983. *Agricultural Development in the Mexican Tropics: Alternatives for the Selva Lacandona Region of Chiapas.*

Rosero-Bixby and Palloni. 1998. *Population and Environment* 20(2): 149–185.

Rudel et al. 2000. *Professional Geographer* 52(3): 386–397.

Sader et al. 1994. *Human Ecology* 22(2): 317–332.

Sagawe. 1991. *Geography* 76(333): 304–314.

Sambrook et al. 1999. *Professional Geographer* 51(1): 25–40.

Schelhas. 1994. *Society and Natural Resources* 7: 67–84.

Schelhas. 1996. *Human Organization* 55(3): 298–306.

Schwartz. 1987. *Journal of Anthropological Research* 43: 163–83.

Sohn et al. 1999. *Photogrammetric Engineering and Remote Sensing* 65(8): 947–958.

Steinberg. 1998. *The Professional Geographer* 50(4): 407–417.

Stevenson. 1989. *Journal of Developing Areas* 24: 59–76.

Stonich and DeWalt. 1996. In *Tropical Deforestation: The Human Dimension.*

Southworth and Tucker. 2001. *Mountain Research and Development* 21(3): 276–283.

Sunderlin and Rodriguez. 1996. *Cattle, Broadleaf Forests, and the Agricultural Modernization Law of Honduras: The Case of Olancho.*

Tole. 2002. *Biodiversity and Conservation* 11: 575–598.
Townsend. 1996. In *Green Guerrillas: Environmental Conflicts and Initiatives in Latin America and the Caribbean: A Reader.*
Utting. 1993. *Trees, People, and Power: Social Dimensions of Deforestation and Forest Protection in Central America.*
de Vos. 1993. *Historical Landscapes: Campeche, Chiapas, Quintana Roo, Tabsco, Yucatan.*
Weis. 2000. *Global Environmental Change* 10: 299–305.

Qualitative Comparative Analysis

Deforestation in the 1980s (37/43)

RANCHING COLON SMALLAG (10—Arizpe, Brothers, Camargo, Harrison, Hernandez, Heckadon-Moreno, Jones, Lobato, Peckenham, Townsend) +
LOGGING RANCHING COLON (7—de Vos, Joly, Fuentes Aguilar, McKay, O'Brien, Schwartz, Utting) +
RANCHING SMALLAG POV (3—Carriere, Schelhas94, Stonich) +
LOGGING pov COLON (4—Halhead, Sader, Geoghegan, Ludeke) +
SMALLAG POV colon (2—Rosero-Bixby, Stevenson) +
LOGGING RANCHING smallag pov (3—Dirzo, Nations, Price) +
logging ranching POV colon (3—Catanese, Lewis, Sagawe) +
RANCHING smallag pov colon (4—Mendosa, Janzen, Parsons, Place)
Contradictories = 6 (Collier, DeWalt, Eyre, Pierce, Sambrook, Rudel)

Deforestation in the 1990s (19/19)

logging pov TENURE SMALLAG (2—Sohn, Weis) +
RANCHING pov TENURE ROADS SMALLAG (3—Carr, Sunderlin, Humphries) +
logging ranching pov ROADS SMALLAG (2—Chomitz, Nygren) +
logging ranching TENURE roads SMALLAG (1—Steinberg) +
logging ranching POV TENURE roads (2—Godoy, Tole) +
logging RANCHING POV tenure roads SMALLAG (2—Faris, Larson) +
LOGGING ranching tenure roads SMALLAG (1—Monaghan) +
LOGGING pov tenure roads SMALLAG (3—Ochoa-Gaona, Geoghegan, Cortes and Restrepo)

Afforestation in the 1990s (9/9)

affpol OUTMIGR smallag NONFARM (1—Klooster) +
affpol PRICEDEC smallag NONFARM (2—Collier, Hecht) +
affpol PRICEDEC OUTMIGR smallag (1—Rudel) +
AFFPOL PRICEDEC OUTMIGR SMALLAG nonfarm (1—Schelhas96) +
AFFPOL pricedec outmigr SMALLAG nonfarm (1—Berti Lungo) +
AFFPOL pricedec OUTMIGR smallag nonfarm (1—Southworth)

South America

Case Studies

Andersen et al. 2002. *The Dynamics of Deforestation and Economic Growth in the Brazilian Amazon.*

Anderson. 1990. *World Development* 18(9): 1191–1205.

Andrade and Ruiz. 1988. *Amazonia Colombiana: Una Aproximacion a la Problematica Ecologica y Social de la Colonizacion del Bosque.* FESCOL, Bogota.

Assies. 1997. *Going Nuts for the Rain Forest: Non-Timber Forest Products, Forest Conservation and Sustainability in Amazonia.*

Becker. 1986. *Spontaneous/induced Rural Settlement in Brazilian Amazonia.*

Bedoya and Klein. 1996. In *Tropical Deforestation: The Human Dimension.*

Behrens et al. 1994. *Human Ecology* 22(3): 279–315.

Bromley. 1981. *Singapore Journal of Tropical Geography* 2(1): 15–26.

Brondizio et al. 1994. *Human Ecology* 22(3): 249–277.

Browder. 1994. *Studies in Comparative International Development* 29(3): 45–69.

Buschbacher et al. 1987. In *Amazonian Rain Forests: Ecosystem Disturbance and Recovery.*

Carneiro. 1979. *Anthropologica* 52: 39–76.

Collins. 1988. *Unseasonal Migrations: The Effects of Rural Labor Scarcity in Peru.*

Coomes et al. 2000. *Ecological Economics* 32: 109–124.

Dale et al. 1997. *Photogrammetric Engineering and Remote Sensing* 59(6): 997–1005.

Echavarria. 1991. *Revista Geografica* 114: 37–53.

Ekstrom. 1979. In *Cultural Transformations and Ethnicity in Modern Ecuador.*

Faminow. 1998. *Cattle, Deforestation, and Development in the Amazon: An Economic, Agronomic and Environmental Perspective.*

Fearnside. 1986. *Ambio* 15(2): 74–81.

Fearnside. 1993. *Ambio* 22(8): 537–545.

Fearnside. 2002. *Deforestation and Land Use in Amazonia.*

Fujisaka et al. 1996. *Agriculture, Ecosystems, and Environment* 59(1/2): 115–125.

Godoy et al. 1996. *The Effects of Economic Development on Neotropical Deforestation: Household and Village Evidence from Amerindians in Bolivia.*

Hecht. 1993. *BioScience* 43(10): 687–695.

Hegen. 1966. *Highways into the Upper Amazon Basin: Pioneer Lands in Southern Colombia, Ecuador, and Northern Peru.*

Hiroaka. 1995. *Global Environmental Change* 5(4): 323–336.

Imbernon. 1999. *Bois et Forets des Tropiques* 259(1): 45–58.

Jones. 1995. In *The Social Causes of Environmental Destruction in Latin America.*

Jordan. 1987. In *Amazonian Rain Forest: Ecosystem Disturbance and Recovery.*

Kaimowitz et al. 1999. *World Development* 27(3): 505–520.

Kaimowitz et al. 2002. In *Deforestation and Land Use in the Amazon.*

Laurance et al. 2001. *Environmental Conservation* 28(4): 305–311.

Laurance et al. 2002. *Journal of Biogeography* 29: 737–748.

Maki et al. 2001. *Environmental Conservation* 28(3): 199–214.

McCracken et al. 2002. In *Deforestation and Land Use in the Amazon.*

Malingreau and Tucker. 1988. *Ambio* 17(1): 49–55.

Mertens et al. 2002. *Agricultural Economics* 27: 269–294.

Moran. 1983. In *The Dilemma of Amazonian Development.*

Moran et al. 2002. In *Deforestation and Land Use in the Amazon.*

Nelson et al. 1987. *International Journal of Remote Sensing* 8(12): 1767–1784.

Nepstad et al. 2001. *Forest Ecology and Management* 154: 395–407.

Ortiz. 1984. In *Frontier Expansion in Amazonia.*

Pfaff. 1999. *Journal of Environmental Economics and Management* 37(2): 26–43.

Pedlowski et al. 1997. *Landscape and Urban Planning* 38(3/4): 149–157.

Pichon et al. 2001. In *Agricultural Technologies and Tropical Deforestation.*

Rudel with Horowitz. 1993. *Tropical Deforestation: Small Farmers and Land Clearing in the Ecuadorian Amazon.*

Rudel et al. 2002. *Annals of the Association of American Geographers* 92(1): 87–102.

Saatchi et al. 1997. *Remote Sensing of Environment* 59(2): 191–202.

Sawyer. 1984. In *Frontier Expansion in Amazonia.*

Schmink and Wood. 1992. *Contested Frontiers in Amazonia.*

Sierra. 2000. *Applied Geography* 20(1): 1–16.

Skole and Tucker. 1993. *Science* 260: 1905–1910.

Southgate et al. 1991. *World Development* 19(9): 1145–1151.

Steininger et al. 2000. *Environmental Conservation* 28(2): 127–134.

Stone et al. 1991. *Forest Ecology and Management* 38: 291–304.

Vina and Cavelier. 1999. *Biotropica* 31(1): 31–36.

Vosti et al. 2002. *Agricultural Intensification by Smallholders in the Western Brazilian Amazon: From Deforestation to Sustainable Land Use.*

Walker et al. 2000. *World Development* 28(4): 683–699.

Walker and Perz. 2002. *International Regional Science Review* 25(2): 169–199.

Weil and Weil. 1983. *Interamerican Economic Affairs* 36: 29–62.

Woodwell et al. 1986. In *The Changing Carbon Cycle: A Global Analysis.*

Wunder. 2000. *The Economics of Deforestation: The Case of Ecuador.*

Young. 1998. In *Nature's Geography: New Lessons for Conservation in Developing Countries.*

Qualitative Comparative Analysis

Deforestation in the 1980s (30/32)

SMALLAG COLON (13—Andrade, Becker, Bedoya, Browder, Carneiro, Collins, Echavarria, Hegen, Moran, Nelson, Ortiz, Sawyer, Weil, Woodwell) +
SMALLAG ROADS (7—Bromley, Fearnside86, Hecht, Malingreau, Pichon, Rudel with Horowitz, Stone) +
RANCHING SMALLAG (2—Eckstrom, Southgate) +
ranching ROADS (2—Young, Pfaff) +
ROADS colon (4—Anderson, Buschbacher, Skole, Vina)
Contradictories = 2 (Coomes, Jordan)

Deforestation in the 1990s (32/32)

SMALLAG ROADS (12—Behrens, Dale, Faminow, Fujisaka, Laurance01, Laurance02, McCracken, Moran02, Pedowski, Saatchi, Vosti, Walker) +
RANCHING ROADS (3—Schmink, Mertens, Nepstad) +
ROADS logging (7—Fearnside02, Imbernon, Kaimowitz99, Kaimowitz02, Maki, Sierra, Walker and Perz) +
RANCHING SMALLAG LOGGING (5—Andersen, Fearnside93, Godoy, Jones, Wunder) +
RANCHING smallag logging (2—Assies, Steininger)

West Africa

Case Studies

Amanor. 1994. *The New Frontier: Farmers' Response to Land Degradation: A West African Study.*
Awung. 1998. *Underlying Causes of Deforestation and Forest Degradation in Cameroon.* Paper given at a workshop on the Underlying Causes of Deforestation and Forest Degradation in Africa. Accra, Ghana. Available at www.wrm.org.uy.
Benneh. 1988. In *Agricultural Expansion and Pioneer Settlement in the Humid Tropics.* Available at www.unu.edu/unupress/unupbooks.
Bertrand. 1983. *Revue Bois et Forets des Tropiques* 202(4): 3–17.
Boateng and Adomako. 1998. *The Underlying Causes of Deforestation: A Case Study of the Tain Tributaries II Forest Reserve and its Surrounding Areas in the Brong Ahafo Region of Ghana.* Paper given at a workshop on the Underlying Causes of Deforestation and Forest Degradation in Africa. Accra, Ghana. Available at www.wrm.org.uy.
Dei. 1990. *Anthropologica* 32: 3–27.
Dorm-Adzobu. 1974. *Bulletin of the Ghana Geographical Association* 16: 45–53.
Fa. 1991. *Conservacion de los Ecosistemas Forestales de Guinea Ecuatorial.*
Fairhead and Leach. 1995. *World Development* 23(6): 1023–1035.

Gilruth et al. 1990. *Photogrammetric Engineering and Remote Sensing* 56(10): 1375–1382.

Hill. 1963. *The Migrant Cocoa Farmers of Southern Ghana.*

Ite and Adams. 1998. *Applied Geography* 18(4): 301–314.

Jones et al. 1991. *Conservacao dos Ecossistemas Florestais na Republica Democratica de Sao Tome e Principe.*

Leach. 1994. *Rainforest Relations: Gender and Resource Use Among the Mende of Gola, Sierra Leone.*

Leach and Fairhead. 2000. *Population and Development Review* 26(1): 17–43.

Martin. 1991. *The Rain Forests of West Africa: Ecology, Threats, and Conservation.*

Mertens and Lambin. 2000. *Annals of the Association of American Geographers* 90(3): 467–494.

Nyerges. 1988. *Swidden Agriculture and the Savannization of Forests in Sierra Lione.*

Nyerges and Green. 2000. *American Anthropologist* 102(2): 271–289.

Oates. 1999. *Myth and Reality in the Rain Forest: How Conservation Strategies Are Failing in West Africa.*

Rubin et al. 1998. *Marine Pollution Bulletin* 37(8–12): 441–449.

Wayi. 1998. *Underlying Causes of Deforestation and Forest Degradation: A Case Study of the Indigenous Ogoni of Nigeria.* Paper given at a workshop on the Underlying Causes of Deforestation and Forest Degradation in Africa. Accra, Ghana. Available at www.wrm.org.uy.

Qualitative Comparative Analysis

Deforestation in the 1980s (12/12)

Topo POP (3—Nyerges, Bertrand, Benneh) +
INTLMKT POP LOGGING (1—Fa) +
logging intlmkt POP (1—Gilruth) +
logging topo INTLMKT (2—Dorm-Adz, Hill) +
LOGGING topo intlmkt (2—Dei, Martin)

Deforestation in the 1990s (10/10)

FORSIZE SMALLAG FIRE (1—Aung) +
URBMKT FORSIZE SMALLAG (4—Mertens and Lambin, Wayi, Ite, Oates) +
urbmkt SMALLAG FIRE (2—Boateng, Nyerges and Green) +
URBMKT forsize smallag fire (1—Rubin et al.)

Central Africa

Case Studies

Berta et al. 1990. *Geocarto International* 4: 57–61.

Doumenge. 1990. *La Conservation des Ecosystemes Forestiers du Zaire.*

Hecketsweiler. 1990. *La Conservation des Ecosystemes Forestiers du Congo.*

Laporte et al. 1995. *International Journal of Remote Sensing* 16(6): 1127–1145.
Malaisse and Binzangi. 1985. *Commonwealth Forestry Review* 64(3): 227–239.
Mamingi et al. 1996. *Spatial Patterns of Deforestation in Cameroon and Zaire.*
Massart et al. 1995. *Photogrammetric Engineering and Remote Sensing* 61(9): 1153–1158.
Mertens et al. 2000. *World Development* 28(6): 983–999.
Peterson. 2000. *Conversations in the Rainforest: Culture, Values, and the Environment in Central Africa.*
Wilks. 1990. *La Conservation des Ecosystemes Forestiers du Gabon.*
Wilkie et al. 1992. *Conservation Biology* 6(4): 1–11.
Wilkie. 1996. In *Tropical Deforestation: the Human Dimension.*
Wilkie and Finn. 1988. *Ecological Modelling* 41: 307–323.
Witte. 1992. *The Ecologist* 22(2): 58–64.
Wunder. 2003. *Oil Wealth and the Fate of the Forests.*

Qualitative Comparative Analysis

Deforestation in the 1980s (9/9)

urbmkt colon fuelwood SMALLAG ROADS POP (2—Peterson, Wilkie and Finn) +
URBMKT colon FUELWOOD SMALLAG ROADS POP (1—Doumenge) +
URBMKT colon FUELWOOD smallag roads POP (1—Malaisse) +
URBMKT colon fuelwood SMALLAG ROADS pop (2—Berta, Mertens) +
URBMKT COLON fuelwood SMALLAG roads pop (1—Massart)

Deforestation in the 1990s (6/6)

logging URBMKT SMALLAG ROADS (2—Wilkie et al., Laporte) +
URBMKT SMALLAG ROADS pop (2—Mamingi, Witte)

East Africa

Case Studies

Abbot and Homewood. 1999. *Journal of Applied Ecology* 36(3): 422–433.
Allen. 1985. *Journal of Development Economics* 19: 59–84.
Bewket. 2002. *Mountain Research and Development* 22(3): 263–269.
Chidumayo. 1989. *Land Degradation and Rehabilitation* 1: 209–216.
Chidumayo. 2002. *Journal of Biogeography* 29: 1619–1626.
Elnagheeb and Bromley. 1994. *Agricultural Economics* 10: 193–200
Green and Sussman. 1990. *Science* 248: 212–215
Hlavka and Strong. 1992. *Journal of Imagery Science and Technology* 36(5): 440–445.
Holden. 1997. *Forum for Development Studies* 1: 117–134.
Holmgren et al. 1994. *Ambio* 23(7): 390–395.

Hudak and Wessman. 2000. *Applied Geography* 20(2): 155–175.
Imbernon. 1999. *Agriculture, Ecosystems and Environment* 76(1): 67–73.
Kaoneka and Solberg. 1997. *Agriculture, Ecosystems, and the Environment* 62: 59–70.
Kull. 1998. *Professional Geographer* 50(2): 163–176.
Medley. 1998. In *Nature's Geography: New Lessons for Conservation in Developing Countries.*
Minde et al. 2001. *Forests, Trees, and Livelihoods* 11: 167–182.
Namaalwa et al. 2001. *International Forestry Review* 3(4): 299–306.
Place and Otsuka. 2000. *Land Economics* 76(2): 233–251.
Place and Otsuka. 2001. *Journal of Environmental Economics and Management* 41: 13–32.
Prins and Kikula. 1996. *Forest Ecology and Management* 84: 263–266.
Reid et al. 2000. *Landscape Ecology* 15: 339–355.
Struhsaker. 1997. *Ecology of an African Rain Forest.*
Tekle and Hedlund. 2000. *Mountain Research and Development* 20(1): 42–51.
Tiffen et al. 1994. *More People, Less Erosion: Environmental Recovery in Kenya.*

Qualitative Comparative Analysis

Deforestation in the 1970s and 1980s (12/12)

urbmkt FUELWOOD POP SMALLAG (1—Tekle and Hedlund) +

high urbmkt POP SMALLAG (5—Medley, Prins, Green, Hlavka-Strong, Struhsaker) +

high urbmkt FUELWOOD SMALLAG (2—Hudak, Elnagheeb and Bromley) +

high urbmkt FUELWOOD POP (1—Abbot) +

high FUELWOOD POP smallag (2—Chidumayo89, Allen)

Deforestation in the 1990s (11/13)

high fuelwood tenure POP SMALLAG (4—Minde, Holden, Imbernon, Place and Otsuka01) +

high URBMKT FUELWOOD TENURE POP SMALLAG (2—Chidumayo02, Place and Otsuka00) +

HIGH urbmkt fuelwood TENURE pop SMALLAG (1—Reid) +

high urbmkt FUELWOOD tenure pop SMALLAG (1—Namaalwa)

Contradictories = 2 (Kaoneka, Bewket)

South Asia

Case Studies

Agrawal. 2000. In *People and Forests: Communities, Institutions, and Governance.*
Alcorn and Molnar. 1996. In *Tropical Deforestation: The Human Dimension.*

Awasthi et al. 2002. *Land Degradation and Development* 13: 495–513.
Bajracharya. 1983. *World Development* 11(2): 1057–1074.
Bowonder. 1982. *International Journal of Environmental Studies* 18: 223–236.
Bowonder et al. 1987. *Environmental Conservation* 14(1): 23–28.
Chaffey et al. 1985. *A Forest Inventory of the Sunderbans, Bangladesh.*
Faminow and Klein. 2000. *Canadian Journal of Agricultural Economics* 48(4): 585–595.
Foster and Rosenzweig. 2003. *The Quarterly Journal of Economics* 118(2): 601–637.
Guha et al. 1984. *Journal of Indian Anthropology and Sociology* 19: 246–253.
Gupta et al. 1981. *Tribal Unrest and Forestry Management in Bihar.*
Jackson et al. 1998. *Mountain Research and Development* 18(3): 193–212.
Kohlin and Parks. 2001. *Land Economics* 7(2): 206–218.
Kothari et al. 1989. *Management of National Parks and Sanctuaries in India: A Status Report.* Indian Institute of Public Administration, New Delhi.
Mukherjee. 1997. *Indian Forester* 124: 460–471.
Rao and Pant. 2001. *Agriculture, Ecosystems and Environment* 86: 113–123.
Rathore et al. 1997. *Journal of Environmental Management* 49: 265–276.
Ravindranath and Hall. 1994. *Ambio* 23(8): 521–523.
Rawat. 1995. *Mountain Research and Development* 15(4): 311–322.
Richards. 1987. *Mountain Research and Development* 7(3): 299–304.
Schreier et al. 1998. *Environmental Management* 18(1): 139–150.
Schweik et al. 1997. *Mountain Research and Development* 17(2): 99–116.
Sen et al. 2002. *Mountain Research and Development* 22(1): 56–62.
Singh. 2002. *Forests, Trees, and People* #46.
Soussan et al. 1995. *The Social Dynamics of Deforestation: a Case Study from Nepal.*
Varughese. 2000. In *People and Forests: Communities, Institutions, and Governance.*

Qualitative Comparative Analysis

Deforestation in the 1980s (12/12)

FUELWOOD POP central (4—Richards, Soussan et al., Bajracharya, Bowonder et al.) +
SMALLAG FUELWOOD POP LOGGING (3—Alcorn, Agrawal, Bowonder) +
smallag FUELWOOD central LOGGING (1—Chaffey) +
SMALLAG fuelwood pop CENTRAL LOGGING (1—Gupta) +
SMALLAG fuelwood pop central logging (2—Kothari, Singh)

Deforestation in the 1990s (12/14)

SMALLAG FUELWOOD pop central (1- Rawat) +
SMALLAG FUELWOOD central logging (3—Schweik, Sen, Rathore)
Contradictories = 2 (Varughese, Rao)

Afforestation in the 1990s (12/14)

logging SMALLAG pop LOCAL (2—Faminow, Mukherjee) +
SMALLAG pop LOCAL fuelwood (2—Schreier, Awasthi) +
LOGGING smallag pop LOCAL (1—Kohlin and Parks) +
LOGGING pop LOCAL fuelwood (1—Ravindranath and Hall) +
logging smallag POP LOCAL FUELWOOD (1—Jackson) +
SMALLAG FUELWOOD POP LOGGING local (1—Foster and Rosenz-
weig)
Contradictories = 2 (Varughese, Rao)

Southeast Asia

Case Studies

Aiken and Leigh. 1992. *Vanishing Rain Forests: The Ecological Transition in Malaysia.*
Aliño. 1993. *Decisions on the Uplands.*
Angelsen. 1995. *World Development* 23(10): 1713–1729.
Bee. 1987. *Depletion of the Forest Resources in the Philippines.*
Berger. 1990. *Malaysia's Forests: A Resource Without a Future?*
Bevis. 1995. *Borneo Log: the Struggle for Sarawak's Forests.*
Colfer and Dudley. 1993. *Shifting Cultivators of Indonesia: Marauders or Managers of the Forest?*
Colfer et al. 1997. In *Beyond Slash and Burn: Building on Indigenous Management of Borneo's Tropical Rain Forests.*
Cook. 1996. In *Tropical Deforestation: The Human Dimension.*
Coxhead et al. 2000. In *Agricultural Technologies and Tropical Deforestation.*
Cropper et al. 2001. *Land Economics* 77(2): 172–186.
Cruz and Cruz. 1990. *ASEAN Economic Bulletin* 7(2): 200–212.
Davis. 1988. In *People of the Tropical Rain Forest.*
de Lang. 2002. *Society and Natural Resources* 15: 483–501.
Dennis, Colfer et al. 2001. In *People Managing Forests: the Links between Human Well Being and Sustainability.*
Eder. 1990. *Population and Environment* 12(2): 99–115.
Fox et al. 1995. *Ambio* 24(6): 328–334.
Fox et al. 2001. *Kluwer International Series in Engineering and Computer Science* 626: 289–308.
Fraser. 1998. In *Human Activities and the Tropical Rainforest: Past, Present, and Possible Future.*

Geddes. 1976. *Migrants of the Mountains: The Cultural Ecology of the Blue Miao of Thailand.*

Heaney et al. 1988. *Vanishing Treasures of the Philippine Rain Forest.*

Hiebert. 1994. *Far Eastern Economic Review* 157(14): 64.

Hofner and Apichatuallop. 1990. In *Keepers of the Forest: Land Management Alternatives in Southeast Asia.*

Holden and Huoslef. 1995. In *Management of Tropical Forests: Towards an Integrated Perspective.*

Imbernon. 1999. *Agriculture, Ecosystems and Environment* 76(1): 61–66.

Indrabudi et al. 1998. *Land Degradation and Development* 9(4): 311–322.

James. 1983. *Economic Development and Cultural Change* 31(3): 571–586.

de Jong. 2001. *Journal of Tropical Forest Science* 13(4): 705–726.

Kartawinata et al. 1984. *Environmentalist* 4: 87–98.

Kummer. 1992. *Agroforestry Systems* 18: 31–46.

Kummer et al. 1994. *Geographical Review* 84(3): 266–276.

Kunstadter and Chapman. 1978. In *Farmers in the Forest: Economic Development and Marginal Agriculture in Northern Thailand.*

Kuntz and Siegert. 1999. *International Journal of Remote Sensing* 20(4): 2835–2853.

Liu et al. 1993. *Forest Ecology and Management* 57(1): 1–16.

McMorrow and Talip. 2001. *Global Environmental Change* 11(3): 217–230.

Muller and Zeller. 2002. *Agricultural Economics* 27: 333–354.

Nibbering. 1999. *Agroforestry Systems* 46(1): 65–82.

Panayotou and Sungsuwan. 1989. *An Econometric Study of the Causes of Tropical Deforestation: The Case of Northeast Thailand.*

Paulson. 1994. *Environmental Conservation* 21(4): 326–331.

Pelzer. 1968. *Journal of Asian Studies* 27(2): 269–279.

Potter and Lee. 1998. *Tree Planting in Indonesia: Trends, Impacts, and Directions.*

Shively and Martinez. 2001. In *Agricultural Technologies and Tropical Deforestation.*

Sikor and Truong. 2002. *Mountain Research and Development* 22(3): 248–255.

Sikor. 2001. *Development and Change* 32(5): 923–949.

Simkins and Wernstedt. 1971. *Philippine Migration: The Settlement of the Digos-Padada Valley, Davao Province.*

Sunderlin et al. 2001. *World Development* 29(5): 767–782.

Tachibana et al. 2001. *Journal of Environmental Economics and Management* 41(1): 44–69.

Taylor et al. 1994. *Geoforum* 25(3): 351–369.

Thapa. 1998. *Singapore Journal of Tropical Geography* 19(1): 71–91.

Vayda and Sahur. 1985. *Indonesia* 39: 93–110.

Wallace. 1995. *Human Organization* 54(2): 182–186.

Walters. 2003. *Environmental Conservation* 30(2): 293–303.
Ward. 1995. *Pacific Viewpoint* 36(1): 73–93.

Qualitative Comparative Analysis

Deforestation in the 1980s (28/33)

LOGGING SMSEDAG pop (1—Liu) +

LOGGING SMSEDAG hills (2—Kummer, Kuntz) +

LOGGING SMSEDAG colon (7—Alino, Bee, Cruz and Cruz, Fraser, Hofner, Kummer et al., Simkins) +

SMSEDAG pop hills (4—Davis, James, Holden, Vayda and Sahur) +

SMSEDAG colon hills (1—Cropper) +

logging POP COLON HILLS (3—Eder, Kunstadter and Chapman, Muller and Zeller) +

LOGGING POP COLON hills (4—Aiken, Berger, Colfer and Dudley, Panayotou)

Contradictories = (5—Geddes, Dennis-Colfer et al., Paulson, Nibbering, Sikor)

Deforestation in the 1990s (21/21)

SMSEDAG intense tenure roads (2—Imbernon, Wallace) +

LOGGING PLANT intense TENURE (3—Colfer et al., Heaney, Potter and Lee) +

SMSEDAG LOGGING PLANT intense roads (3—Cook, Kuntz, Sunderlin) +

smsedag LOGGING intense TENURE ROADS (2—Bevis, McMorrow) +

SMSEDAG PLANT logging intense tenure (1—Hiebert) +

SMSEDAG LOGGING plant INTENSE TENURE roads (1—Thapa) +

logging SMSEDAG plant intense roads (2—Angelsen, Coxhead) +

logging smsedag plant intense tenure ROADS (1—Indrabucki)

Afforestation (51/54)

INTENSE SMSEDAG HILLS logging (4—Shively, Tachibana et al., Fox et al.01, Sikor and Truong2) +

INTENSE HILLS ROADS POP logging (1—Fox et al.95) +

INTENSE SMSEDAG HILLS roads pop (2—Muller and Zeller, Walters)

INTENSE SMSEDAG log roads pop (1—de Jong)

intense smsedag hills log roads POP (1—Dennis and Colfer)

Contradictory = 3 (Nibbering, Paulson, Sikor)

Notes

Chapter One

1. During the 1980s, approximately 9.2 million hectares of tropical forest disappeared each year; between 1990 and 2000, the deforestation rate slowed only slightly, to 8.6 million hectares annually (FAO, 2001). The decline from the 1980s to the 1990s may be more apparent than real. Because the FAO figures represent net changes in forest cover, some of the decline in deforestation rates most likely represents increases in the extent of forest plantations in tropical countries rather than a decline in the rate at which primary rain forests are disappearing (Mathews, 2001). The FAO estimates for both decades are open to question. Using highly reliable remote sensing measurement procedures, scientists associated with the European Union's TREES program contend that tropical deforestation during the 1990s was 23 percent lower than the FAO estimates (Achard et al., 2002). Because the TREES scientists did not compute a comparable deforestation figure for the 1980s, we must rely on the FAO measures for estimates of the trends over time in deforestation rates.
2. For a critical review of the methods used in FAO's surveys of tropical forest cover, see Wunder (2003).
3. FAO's recalculation of 1990 forest areas produced 35 gains in forest area and 30 losses in forest area among the 65 countries with significant

amounts of tropical forest. Outside of Africa, only Mexico and Myanmar saw significant gains in forest area.

4. Noise represents random error introduced through the use of faulty measures. Bias involves nonrandom error, and the patterned nature of the error skews estimates in a systematic way.

5. The rates of change in forest cover were calculated by summing the areas in forest for all of the countries in a region for 1980 and 1990, computing the change in forest cover during the 1980s, and then dividing the change by the base areas in forest in 1980.

6. The countries included in each of these regions are listed in table 2.2.

7. Globalization has two common meanings. First, it refers to the geographic expansion of markets from internal to international in scale. Second, it can also refer to the creation of institutions with global reach, such as the International Monetary Fund. The term is used in both ways in this book.

8. Conjoint causation could be represented in variable-oriented statistical analyses by interaction terms that estimate the joint effects of two or more independent variables on a dependent variable. Although in concept interaction terms can represent situations of conjoint causation, in practice most multivariate analysts rarely construct models with interaction terms, preferring simpler linear models of independent variables with additive effects (Ragin, 1987:67).

Chapter Two

1. It is important to distinguish between organizations and institutions. Institutions are "the rules of game in a society" (North, 1990:3). Organizations are networks of people who work together formally toward a common goal.

2. Personal communication, Christine Padoch, February 2001.

3. There are other differences between the two studies. One of the most notable involves the use of the distinction between proximate and underlying causes. In practice, I have had difficulties deciding whether to classify a particular driving force (e.g., population growth) as a proximate or as an underlying cause, so I have avoided using this binary. Most of the causal forces identified in the qualitative comparative analyses in *Tropical Forests* would be considered proximate causes in the Geist and Lambin analytical framework.

4. This method can be extended to data that do not limit membership in a set to an either/or proposition. In these analyses, there are degrees of membership in a set, creating, in effect, "fuzzy sets" (Ragin, 2000). The protocols for carrying out fuzzy set analyses are outlined in Ragin (2000).

5. The truth table in table 2.1 has been simplified for the purposes of illustration. It contains fewer studies of Ecuador, 8 instead of 20, and fewer possible causes, 5 instead of 20, than the truth tables that were used for

the regional analyses of South America. By extension, the minimized versions of this truth table present an oversimplified configuration of causes for Ecuadorian deforestation between 1980 and 2000.

6. Some authors write multiple articles out of a single piece of field research. To avoid counting these articles as more than one study, they enter the database as only one case. An author's work enters the database several different times when she or he has carried out several case studies in different locales or at different times.

7. For a fuller exposition of these points, see Ragin (1987:85–164).

8. Personal communication, Billie Lee Turner, November 2002.

Chapter Three

1. The source for all of the maps of regional tropical forests in this book is a geographic information system (GIS) database maintained by the World Conservation Monitoring Centre in Cambridge, England. I obtained the data for the maps in 2001. Their construction posed special challenges. The high degree of fragmentation in the forests has sometimes made it difficult to distinguish the different classes of forest from one another in the small islands of forest that dot the maps. To make the maps more legible under these circumstances, I have limited the number of forest classes, and I have used shades of solid fill to denote each class of forest.

2. Biodiversity losses from coffee cultivation have probably varied by region. In Brazil, where cultivators planted coffee in open fields, the losses of flora and fauna were probably more severe than in the Central American hill region, where cultivators usually grew coffee in shaded fields.

3. The colonization zones were usually sparsely populated, but they were not empty. They contained lowland Amerindian populations who regarded the colonists as intruders.

4. Brothers (1997) notes the contrast in the environmental history of the two islands.

5. In this particular instance, a controversy arose over whether or not the teak plantation had been properly certified as grown in a sustainable fashion. For a description of the controversy, see www.treemail.nl/teakscan.dal.

6. Table 3.2 reports the first of three qualitative comparative analyses of afforestation. In addition to this analysis of afforestation in Central America, there are QCAs of afforestation in south Asia and in Southeast Asia later in the book. The Central American QCA was carried out in a manner that was somewhat different from that used for the south Asian and in Southeast Asian QCAs. Because the afforestation analysis in Central America involved a set of variables that was completely different from those in the deforestation analysis of Central America, I created two data sets for the region—one for deforestation and another for afforestation. In the analyses for south Asia and Southeast Asia, I could use quite similar

sets of variables for the afforestation and for the deforestation analyses, so each region has just one data set. In each region, I minimized the data for deforestation and afforestation, producing, in this way, several QCAs for each region. This difference in analytical procedure between the Central American and Asian regions does not appear to affect the substantive conclusions from the analyses.

Chapter Four

1. More precisely, the Amazon basin contains 45.7 percent of the world's tropical forests according to the FAO's most recent compilation of data. This estimate defines the Amazon somewhat expansively, including the entire Guiana region north of the Amazon and east of the Orinoco River; it also defines tropical forests in an inclusive way; the denominator in this calculation, the worldwide extent of tropical forests, includes the dry tropical forests of south central Africa.
2. Field notes, Ecuador, 2001; personal communication, David Hughes, 2001.
3. The Gini coefficient for the Brazilian income distribution in 1997 is .591. Of the populous countries in the world, only South Africa has a more skewed income distribution.
4. Personal communication, Emilio Moran, November 2002.
5. The actual extent of the cleared land depends of course on the degree to which the Brazilian state enforces the law requiring that all landowners leave 50 percent of their lands in forest.
6. Expenditures for beef and soybean products, unlike for many other agricultural products, tend to become a larger proportion of household expenditures as income increases. An income elasticity of 0.6 for beef means that for every 1 dollar increase in income, a household will spend an additional 60 cents on beef products.
7. For maps indicating the spatial distribution of Amerindian reserves in the Brazilian Amazon, see the FUNAI website (FUNAI, 2001).

Chapter Five

1. Fairhead and Leach (1998) argue that colonial foresters systematically overestimated the original extent of West African forests. These inflated estimates then became the baseline from which more recent analysts have calculated the extent of deforestation in the modern era. In this way, the colonial era biases have contributed to overestimates of contemporary deforestation. Although this argument seems reasonable when applied to the first estimates of regional deforestation, it does not apply to the more recent, short-term estimates of deforestation during the 1980s and 1990s that I discuss in the text.
2. By West Africa, I mean the following countries: Benin, Ivory Coast,

Equatorial Guinea, the Gambia, Ghana, Guinea, Guinea-Bissau, Liberia, Nigeria, Sao Tome and Principe, Senegal, Sierra Leone, and Togo. The definition of West Africa used here groups Cameroon with West Africa rather than central Africa. In truth, Cameroon is a hybrid in terms of its processes of deforestation. The rain forests of the coastal regions exhibit processes of deforestation that resemble those that occur in other West African states. The far southeastern corner of the country, in the Congo River watershed, exhibits a deforestation process more typical of the central African countries within the same watershed. Because most studies of deforestation in Cameroon have been done in the coastal region, loosely defined, I have grouped it with the West African countries.

3. One could also object to this policy emphasis on moral grounds in that it takes resources away from the poorest of the poor and treats them in a punitive way. (See Brechin et al., 2002.)

4. Tenure issues may make it particularly difficult for smallholders to establish tree plantations in West Africa. Typically, the community owns the land. Individuals own the trees that they plant but not the land that they cultivate. Because the ownership of trees in many instances strengthens the individual's claim to the land on which they grow, elders have in at least several instances prevented cultivators from planting trees. Elders have prevented the young from planting trees in northern Ivory Coast (DePlaen, 2001), and local residents have prevented migrants from planting tree crops in southern Ivory Coast (Martin, 1991).

5. Farm-bush fallows vary from one another in the type of secondary vegetation that predominates in them. Strategies for rehabilitating old cocoa lands work to different degrees in the different types of fallows. (For details, see Ruf, 2001:301–302.)

6. Wunder (2001) describes a similar developmental sequence for smallholder banana growers in Ecuador, in which investments in the land, in particular in irrigation systems, sedentarizes the growers.

7. The political repercussions of these new projects can include heightened conflict between villagers as they struggle for control over the resources provided by outside donors. Schroeder (1999) describes the conflicts that emerged in villages of the Gambia after the U.S. Agency for International Development (USAID) and the European Community sponsored woodlot and agroforestry projects during the 1980s.

8. Almost all West African countries would qualify for debt relief from the IMF and the World Bank.

Chapter Six

1. Place names in central and East Africa changed with decolonization, and these changes can create confusion in a reader unfamiliar with the name changes. To reduce confusion about the time and place under discussion in the text, I will always refer to a place by the name it had during the

period under discussion. For example, events before 1960 in the current Democratic Republic of the Congo occurred in the Belgian Congo, but events during the 1970s occurred in Zaire.

2. This macroeconomic process acquired the name Dutch disease because the first analyzed example of the process traced out the effects of a boom in natural gas production in the Netherlands on the other sectors of the Dutch economy. See Corden and Neary (1982).

3. The type of woodlands surrounding a city affects the size of the cleared area. Open, miombo woodlands surround Lubumbashi, and they do not provide as much wood for urban markets as do the more humid, closed forests around, for example, Pointe-Noire in the Congo. For this reason the cutover zone around Pointe Noire is smaller (a 20-km radius) than it is around Lubumbashi and other cities in drier, less wooded environments (Malaisse and Binzangi, 1985:229).

4. Rates of HIV infection for adults, aged 15 to 49, ranged from 4.2 percent in Gabon to 13.8 percent in the Central African Republic in 2000 (World Bank, 2001).

Chapter Seven

1. In "East Africa," I include the following countries: Burundi, Ethiopia, Kenya, Madagascar, Malawi, Mozambique, Rwanda, Sudan, Tanzania, Zambia, and Zimbabwe.

2. East Africa resembles the American Southwest in the way elevation influences the landscape. Both regions contain humid "sky islands" that support closed forests. In both places, rainfall declines and vegetation becomes more sparse at lower elevations. The types of vegetation at lower elevations differ. East Africa contains extensive dry forests at these elevations, whereas the American Southwest features desert landscapes.

3. Without aerial photographs to corroborate this claim, it remains only a reasonable conjecture about forest-cover trends in the region before 1970.

4. There are two case studies of deforestation in the 1980s that indicate the importance of the TENURE variable, but they match up very closely with other case studies that do not indicate TENURE as a driving force. To achieve parsimony, the TENURE term was dropped from these causal configurations (which include POP and SMALLAG).

5. A weak state has three distinguishing characteristics: (1) little legitimacy, (2) inability to mobilize people for actions such as paying taxes that deliver a collective benefit, and (3) difficulty obtaining popular compliance with government directives such as newly demarcated boundaries of forest reserves (Migdal, 1988).

Chapter Eight

1. Nancy Peluso (1992) recounts a similar history of colonial control over forests and peasant resistance to colonial rulers in a study of forests in Dutch-controlled Java during the early twentieth century.

2. In table 8.2, the smaller number (4) of studies accounted for in the causal configurations of part B compared to the larger number in part A (11) reflects the smaller number of studies from the 1990s that reported deforestation.

3. China has more extensive tree plantations than India, even though it plants at a lower rate (1,154,000 per annum), because it began creating tree plantations earlier than India, in the 1960s instead of the 1980s.

4. This generalization applies only to countries with humid climates, where the spontaneous growth of forests is possible. Plantations constitute 100% of the woodland in some Middle Eastern countries such as Kuwait, where arid climates make the natural growth of forests impossible.

5. In all likelihood, there is a curvilinear relationship between the size of forest user groups and the efficacy of their operation. Small size makes it difficult to create a large enough group of people to manage a forest; however, because of diverging interests and failures of communication, large size, beyond a certain point, can make it impossible to create an effective management team (Agrawal, 2000).

Chapter Nine

1. Some readers may question the decision to include Papua New Guinea (PNG) in Southeast Asia. Although the country's vegetation and its people differ dramatically from those of the rest of insular Southeast Asia, similarities in the physical geography and in recent patterns of forest exploitation by Southeast Asian logging firms argue for PNG's inclusion in the region. Certainly PNG is an outlier in the Southeast Asian analysis in the same way that Madagascar is an outlier in the East African analysis.

2. In some instances, these reports may not be reliable. For example, the political desirability of reversing the decline of forests in Vietnam may have affected local reports of forest-cover change in ways that make the national 1990 to 2000 estimates of forest-cover change unreliable. It is, however, true that major increases in tree planting occurred in Vietnam during the 1990s.

3. Dipterocarps dominate to a much greater degree west of Wallace's line than they do east of his line (Collins, Sayer, and Whitmore, 1991:141). Wallace's line runs between the islands of Lombok and Bali in the southern Sunda Isles and between Kalimantan and Sulawesi in the northern Sundas.

4. Only the Indonesian program continued to operate in the 1990s, and after 1998 it resettled people only on lands that had already been cleared for some other purpose (Sunderlin, 1999). The outbreak of interethnic conflict on Kalimantan between native Dayaks and Madurese transmigrants in 2001 raised further questions about the future of this program.

5. One might argue that increasing the yields from rubber trees would just make smallholders more attractive targets for a takeover attempt by a corporation. Although this kind of takeover is certainly conceivable, the relative absence of large-scale rubber plantations in Southeast Asia probably makes the likelihood of this type of takeover attempt fairly low.

Chapter Ten

1. If, as anticipated, the Chinese government extends their road system into the uplands of Laos, Thailand, and Vietnam, the improved access to Chinese markets should indirectly encourage additional agricultural intensification on roadside lands and additional afforestation in the uplands.
2. As Philip Fearnside (2003) has pointed out, purchasing concessions in remote rural areas would also maximize the protected area of tropical forests because the cost per hectare of a concession in a remote rural area would be considerably less than the cost per hectare of a concession in a frontier area.
3. The conditions that facilitate collective action to protect the forests are much more numerous than this sentence would suggest. Poteete and Ostrom (2002) provide a useful list of the conditions that affect the likelihood of community-based forest management. For a more general statement of the conditions that facilitate collective action by communities to manage natural resources, see Ostrom (1999).
4. Personal communication, Arild Angelsen, April 2002.

References

Abbot, Joanne, and K. Homewood. 1999. "A History of Change: Causes of Miombo Woodland Decline in a Protected Area of Malawi." *Journal of Applied Ecology* 36(3): 422–433.

Abdullah, Ibrahim, and Patrick Muana. 1998. "The Revolutionary United Front of Sierra Leone." Pp. 172–194 in C. Clapham (ed.). *African Guerrillas*. Bloomington, Ind.: Indiana University Press.

Abelson, Reed. 2000. "Preserving the Forest by Leasing the Trees." *New York Times*, September 24, Business Section.

Achard, F., H. Eva, A. Glinni, P. Mayaux, T. Richards, and H. Stibig. 1998. "Identification of Deforestation Hot Spot Areas in the Humid Tropics." *TREES Publication Series B*. Research Report #4. Ispra, Italy: European Commission–Joint Research Centre.

Achard, F., H. Eva, H. Stibig, P. Mayaux, J. Gallego, T. Richards, and J. P. Malingreau. 2002. "Determination of Deforestation Rates of the World's Humid Tropical Forests." *Science* 297: 999–1002.

Adams, Jonathan S., and Thomas S. McShane. 1992. *The Myth of Wild Africa: Conservation Without Illusion*. New York: W. W. Norton.

Adas, Michael. 1998. *State, Market, and Peasant in Colonial South and Southeast Asia*. Brookfield, Vt.: Ashgate.

Agrawal, Arun. 2000. "Small Is Beautiful, but Is Larger Better? Forest Management Institutions in the Kumaon Himalaya, India." Pp. 57–86 in C. Gibson, M. McKean, and E. Ostrom (eds.). *People and Forests: Communities, Institutions, and Governance*. Cambridge, Mass.: MIT Press.

————. 2001. "Environmentality: Rethinking the Politics of Nature." Presentation to the Ecology and Culture Seminar. New York: Columbia University.

Alexander's Gas and Oil Connection. 1998. "ARCO Outlines Plans for Block 10 in Ecuadorian Jungle." *Alexander's Gas and Oil Connection* 3(24). Accessed at www.gasandoil.com/goc/company/cnl84281.htm.

Allan, William. 1965. *The African Husbandman*. Edinburgh: Oliver and Boyd.

Allen, Julia. 1985. "Wood Energy and Preservation of Woodlands in Semi-Arid Developing Countries: The Case of Dodoma, Tanzania." *Journal of Development Economics* 19(1): 59–84.

Alves, Diogenes. 2002. "An Analysis of the Geographical Patterns of Deforestation in the Brazilian Amazon in the Period 1991–1996." Pp. 95–106 in Charles Wood and Roberto Porro (eds.). *Deforestation and Land Use in the Amazon*. Gainesville, Fla.: University Press of Florida.

Amacher, Gregory, W. Cruz, D. Grebner, and Wm. Hyde. 1998. "Environmental Motivations for Migration: Population Pressure, Poverty, and Deforestation in the Philippines." *Land Economics* 74(1): 92–101.

Amanor, Kojo Sebastian. 1994. *The New Frontier: Farmers' Response to Land Degradation—A West African Study*. Atlantic Highlands, N.J.: Zed Books.

Andersen, Lykke, C. Granger, E. Reis, D. Weinhold, and S. Wunder. 2002. *The Dynamics of Deforestation and Economic Growth in the Brazilian Amazon*. Cambridge, England: Cambridge University Press.

Anderson, Robert, and W. Huber. 1988. *The Hour of the Fox: Tropical Forests, the World Bank, and Indigenous People in Central India*. Seattle, Wash.: University of Washington Press.

Angelsen, A., and D. Kaimowitz (eds.). 2001. *Agricultural Technologies and Tropical Deforestation*. New York: CAB International.

Angelsen, Arild. 1999. "Agricultural Expansion and Deforestation: Modeling the Impact of Population, Market Forces, and Property Rights." *Journal of Development Economics* 58: 185–218.

Arnold, Michael, G. Kohlin, R. Persson, and G. Shepherd. 2003. "Fuelwood Revisited: What Has Changed in the Last Decade." Center for International Forestry Research (CIFOR) Occasional Paper # 39. Bogor, Indonesia.

Atyi, Richard, and M. Simula. 2002. "Forest Certification: Pending Challenges for Tropical Timber." International Tropical Timber Workshop on Comparability and Equivalence of Forest Equivalence Schemes, Kuala Lumpur, Malaysia, May.

Bailey, Robert C. 1996. "Promoting Biodiversity and Empowering Local People in Central African Forests." Pp. 316–341 in Leslie E. Sponsel, Thomas Headland, and Robert Bailey (eds.). *Tropical Deforestation: The Human Dimension*. New York: Columbia University Press.

Bailey, Robert C., and R. Aunger. 1995. "Sexuality, Infertility, and Sexually Transmitted Disease Among Farmers and Foragers in Central Africa." Pp. 195–222 in P. Abramson and S. D. Pinkerdam (eds.). *Sexual Nature/Sexual Culture*. Chicago: University of Chicago Press.

Bajracharya, Deepak. 1983. "Fuel, Food, or Forest? Dilemmas in a Nepali Village." *World Development* 11(2): 1057–1074.

Balmford, Andrew, A. Bruner, P. Cooper, R. Costanza, S. Farber, R. Green, M. Jenkins, P. Jefferiss, V. Jessamy, J. Madden, K. Munro, N. Myers, S. Naeem, J. Paavola, M. Rayment, S. Rosendo, J. Roughgarden, and K. Trumper. 2002. "Economic Reasons for Conserving Wild Nature." *Science* 297: 950–953.

Balmford, Andrew, J. Moore, T. Brooks, N. Burgess, L. Hansen, P. Williams, and C. Rahbek. 2001. "Conservation Conflicts Across Africa." *Science* 291: 2616–2619.

Banana, Abwoli, and W. Gombya-Ssembajjwe. 2000. "Successful Forest Management: The Importance of Security of Tenure and Rule Enforcement in Ugandan Forests." Pp. 87–98 in Clark Gibson, Margaret McKean, and Elinor Ostrom (eds.). *People and Forests: Communities, Institutions, and Governance*. Cambridge, Mass.: MIT Press.

Barbier, Edward B. 1993. "Introduction: Economics and Ecology—The Next Frontier." Pp. 1–10 in E. Barbier (ed.). *Economics and Ecology: New Frontiers and Sustainable Development*. London: Chapman and Hall.

Barham, Bradford, J. P. Chavas, and O. Coomes. 1998. "Sunk Costs and the Natural Resource Extraction Sector: Analytical Models and Historical Examples of Hysteresis and Strategic Behavior in the Americas." *Land Economics* 74(4): 429–448.

Barham, Bradford, and O. Coomes. 1996. *Prosperity's Promise: The Amazon Rubber Boom and Distorted Economic Development*. Boulder, Colo.: Westview Press.

Barnes, R. F. W., and S. Lahm. 1997. "An Ecological Perspective on Human Densities in the Central African Forests." *Journal of Applied Ecology* 34: 245–260.

Barraclough, Solon, and Krishna Ghimire. 1995. *Forests and Livelihoods: The Social Dynamics of Deforestation in Developing Countries*. London: MacMillan Press.

———. 1996. "Deforestation in Tanzania: Beyond Simplistic Generalizations." *The Ecologist* 26(3): 104–109.

———. 2000. *Agricultural Expansion and Tropical Deforestation: Poverty, International Trade, and Land Use*. London: Earthscan.

Barrett, Christopher, and P. Arcese. 1995. "Are Integrated Conservation and Development Projects Sustainable? On the Conservation of Large Mammals in Sub-Saharan Africa." *World Development* 25: 1073–1084.

Barros, Ana, and C. Uhl. 1995. "Logging Along the Amazon River and Estuary: Patterns, Problems, and Potential." *Forest Ecology and Management* 77: 87–105.

Bassett, Thomas. 2001. "Patrimony and Development: Territories in Northern Cote D'Ivoire." Paper presented at the meetings of the Association of American Geographers, New York.

Bates, Diane. 2001. "Tourism and Economic Development: The Caribbean Experience." Unpublished paper, Department of Sociology, Sam Houston State University, Huntsville, Tex.

Bates, Henry Walter. 1969 (1864). *The Naturalist on the River Amazons*. New York: E. P. Dutton.

Belsky, Jill M. 1999. "Misrepresenting Communities: The Politics of Community-Based Rural Ecotourism in Gales Point Manatee, Belize." *Rural Sociology* 64: 641–666.

Benhin, J. K. A., and E. Barbier. 2001. "The Effects of the Structural Adjustment Program on Deforestation in Ghana." *Agricultural and Resource Economics Review* 30(1): 66–80.

Bergad, Laird. 1978. "Agrarian History of Puerto Rico, 1870–1930." *Latin American Research Review* 13(3): 63–94.

Berry, Sara. 1993. *No Condition Is Permanent: The Social Dynamics of Agrarian Change in Sub-Saharan Africa*. Madison, Wisc.: University of Wisconsin Press.

Berti Lungo, Carlo. 1999. "Transformaciones Recientes en la Industria y La Politica Forestal Costarricense y sus Implicaciones para el Desarrollo de los Bosques Secundarios." Tesis Magister Scientiae, CATIE, Turrialba, Costa Rica.

Bewket, Woldeamlak. 2002. "Land Cover Dynamics Since the 1950s in Chemoga Watershed, Blue Nile Basin, Ethiopia." *Mountain Research and Development* 22(3): 263–269.

Bhat, D. M., K. S. Murali, and N. H. Ravindranath. 2001. "Formation and Recovery of Secondary Forests in India: A Particular Reference to Western Ghats in South India." *Journal of Tropical Forest Science* 13(4): 601–620.

Bilsborrow, Richard, and H. Okoth-Ogendo. 1992. "Population Driven Changes in Land Use in Developing Countries." *Ambio* 21: 37–45.

Blair, H. W. 1996. "Democracy, Equity, and Common Property Resource Management in the Indian Subcontinent." *Development and Change* 27(3): 475–499.

Bluffstone, Randall. 1995. "The Effect of Labor Market Performance on Deforestation in Developing Countries Under Open Access: An Example from Rural Nepal. *Journal of Environmental Economics and Management* 29: 42–63.

Bowonder, B., S. Prasad, N. Unni. 1987. "Deforestation Around Urban Centers in India." *Environmental Conservation* 14(1): 23–28.

Brady, Nyle C. 1988. "International Development and the Protection of Biological Diversity." Pp. 409–418 in E. O. Wilson (ed.). *Biodiversity*. Washington, D.C.: National Academy Press.

Brandon, Katrina. 1998. "Perils to Parks: The Social Context of Threats." Pp. 415–439 in K. Brandon, K. Redford, and S. Sanderson (eds.). *Parks in Peril: People, Politics, and Protected Areas*. Washington, D.C.: Island Press.

———. 2000. "Parks in Peril: The Conservationist View." Presentation, meetings of the Rural Sociological Society. Washington, D.C.

Brechin, Steven, P. Wilshusen, C. Fortwangler, and P. West. 2002. "Beyond the Square Wheel: Toward a More Comprehensive Understanding of Biodiversity Conservation as a Social and Political Process." *Society and Natural Resources* 15(1): 41–64.

Brondizio, Eduardo, E. Moran, P. Mausel, and Y. Wu. 1994. "Land Use Change in the Amazon Estuary: Patterns of Caboclo Settlement and Landscape Management." *Human Ecology* 22(3): 249–277.

Brondizio, Eduardo, and A. Siqueira. 1997. "From Extractivists to Forest Farmers: Changing Concepts of Caboclo Agroforestry in the Amazon Estuary." *Research in Economic Anthropology* 18: 233–279.

Brookfield, Harold. 1994. *Transformation with Industrialization in Peninsular Malaysia.* New York: Oxford University Press.

Brothers, T. 1997. "Deforestation in the Dominican Republic: A Village Level View." *Environmental Conservation* 24(3): 213–223.

Browder, John, and B. Godfrey. 1997. *Rainforest Cities: Urbanization, Development, and Globalization of the Brazilian Amazon.* New York: Columbia University Press.

Bruner, Aaron G., R. Gullison, R. Rice, and Gustavo da Fonseca. 2001. "Effectiveness of Parks in Protecting Tropical Biodiversity." *Science* 291: 125–128.

Bryant, D., D. Nielson, and L. Tangley. 1997. *The Last Frontier Forests: Ecosystems and Economies on the Edge.* Washington: World Resources Institute.

Carneiro, Robert. 1970. "A Theory of the Origin of the State." *Science* 169: 733–38.

Carr, Archie. 1967 (1955). *The Windward Road: Adventures of a Naturalist on Remote Caribbean Shores.* New York: Alfred A. Knopf.

Casson, Anne. 2000. "The Hesitant Boom: Indonesia's Oil Palm Sub-sector in an Era of Economic Crisis and Political Change." Occasional Paper #29. Center for International Forestry Research (CIFOR), Bogor, Indonesia.

Castro, Alfonso P. 1988. "Southern Mount Kenya and Colonial Forest Conflicts." Pp. 33–55 in J. Richards and J. Tucker (eds.). *World Deforestation in the Twentieth Century.* Durham, N.C.: Duke University Press.

———. 1996. "The Political Economy of Colonial Farm Forestry in Kenya: The View from Kirinyaga." Pp. 122–143 in L. Sponsel, T. Headland, and R. Bailey (eds.). *Tropical Deforestation: The Human Dimension.* New York: Columbia University Press.

Catanese, A. V., and R. Perlack. 1993. "Reforestation in Haiti." *Canadian Journal of Development Studies* 14(1): 59–72.

Cavendish, William. 2000. "Empirical Regularities in the Poverty–Environment Relationship of Rural Households: Evidence from Zimbabwe." Unpublished paper.

Central Intelligence Agency (CIA). 2001. *World Factbook.* Available at www.cia.gov/cia/publications/factbook/index.html.

Child, Brian. 2000. "Free Market Conservation in Southern Africa: Using Wildlife to Empower Rural Communities." Presentation, Department of Human Ecology, Rutgers University, New Brunswick, N.J., December.

Cleary, David. 2001. "Towards an Environmental History of the Amazon: From Prehistory to the Nineteenth Century." *Latin American Research Review* 36(2): 65–96.

Cochrane, Mark A., A. Alencar, M. Schulze, C. Souza, D. Nepstad, P. Lefebvre, and E. Davidson. 1999. "Positive Feedbacks in the Fire Dynamic of Closed Canopy Tropical Forests." *Science* 284: 1832–1834.

Colfer, Carol, N. Peluso, and C. Chung. 1997. *Beyond Slash and Burn: Building on Indigenous Management of Borneo's Tropical Rain Forests.* Bronx, New York: The New York Botanical Garden.

Colfer, Carol, J. Pierce, and Yvonne Byron (eds.). 2001. *People Managing Forests: The Links Between Human Well-Being and Sustainability.* Washington, D.C.: Resources for the Future.

Collier, George, D. Mountjoy, and R. Nigh. 1994. "Peasant Agriculture and Global Change: A Maya Response to Energy Development in Southeastern Mexico." *Bioscience* 44(6): 398–407.

Collins, N. Mark, J. Sayer, and T. Whitmore (eds.). 1991. *The Conservation Atlas of Tropical Forests: Asia and the Pacific.* New York: Simon and Schuster.

Corden, W. M., and J. P. Neary. 1982. "Booming Sector and Deindustrialization in a Small Open Economy." *Economic Journal* 92: 825–848.

Coucletis, Helen. 2002. "Why I No Longer Work with Agents." Pp. 3–5 in D. Parker, T. Berger, and S. Manson (eds.). *Agent-Based Models of Land-Use and Land-Cover Change.* LUCC Series Report No. 6. Louvain-la-Neuve, Belgium.

Cour, Jean-Marie, and Serge Snrech (eds.). 1998. *Preparing for the Future: A Vision of West Africa in the Year 2020: West Africa Long-Term Perspective Study.* Paris: Organization for Economic Cooperation and Development.

Cowell, Adrian. 1990. *The Decade of Destruction: The Crusade to Save the Amazon Rain Forest.* New York: Henry Holt.

Cropper, Maureen, J. Puri, and C. Griffiths. 2001. "Predicting the Location of Deforestation: The Role of Roads and Protected Areas in North Thailand." *Land Economics* 77(2): 172–186.

Cruz, Maria C. 2000. "Population Pressure, Poverty, and Deforestation: Philippines Case Study." Paper presented at the Asian Population Network Workshop on Population and Environment, Penang, Malaysia.

Dalton, E. J. T. 1959 (1845). "An Excursion up the Subansiri." Pp. 136–164 in Verrier Elwin (ed.). *India's North-East Frontier in the Nineteenth Century.* Oxford: Oxford University Press.

Dauvergne, Peter. 1997. *Shadows in the Forest: Japan and the Politics of Timber in Southeast Asia.* Cambridge, Mass.: MIT Press.

———. 1998. "Globalization and Deforestation in the Asia–Pacific." *Environmental Politics* 7(4): 114–135.

Deacon, Robert T. 1994. "Deforestation and the Rule of Law in a Cross-Section of Countries." *Land Economics* 70(4): 414–30.

Dean, Warren. 1995. *Between Broadax and Firebrand: The Destruction of the Brazilian Atlantic Forest.* Berkeley, Calif.: University of California Press.

de Jong, Wil. 2001. "The Impact of Rubber on the Forest Landscape in Borneo." Pp. 367–381 in A. Angelsen and D. Kaimowitz (eds.). *Agricultural Technologies and Tropical Deforestation.* New York: CAB International.

de Lang, Claudio. 2002. "Deforestation in Northern Thailand: The Result of Hmong Farming Practices or Thai Development Strategies?" *Society and Natural Resources* 15(6): 483–502.

Denevan, William (ed.). 1992. The Native Population of the Americas in 1492. Second edition. Madison, Wisc.: University of Wisconsin Press.

Dennis, R., A. Hoffmann, G. Applegate, G. von Gemmingen, and K. Kartawinata. 2001. "Large-Scale Fire: Creator and Destroyer of Secondary Forests in Western Indonesia." *Journal of Tropical Forest Science* 13(4): 786–799.

De Plaen, Renaud. 2001. "Contracting Malaria in the Paddies: A Farming Systems Approach to Malaria in Northern Cote D'Ivoire." Ph.D. Dissertation. Department of Geography, Rutgers University, New Brunswick, N.J.

Dewees, Peter. 1994. "Social and Economical Aspects of Miombo Woodland Management in Southern Africa: Options and Opportunities for Research." Center for International Forestry Research (CIFOR) Occasional Paper #2. Bogor, Indonesia.

Dorm-Adzobu, C. 1974. "The Impact of Migrant Ewe Cocoa Farmers in Buem, the Volta Region of Ghana." *Bulletin of the Ghana Geographical Association* 16: 45–53.

Dove, Michael. 1993. "A Revisionist View of Tropical Deforestation and Development." *Environmental Conservation* 20(1): 17–24, 56.

Downton, Mary. 1995. "Measuring Tropical Deforestation: Development of the Methods." *Environmental Conservation* 22(3): 229–240.

Draulans, Dirk, and E. Van Krunkelsven. 2002. "The Impact of War on Forest Areas in the Democratic Republic of the Congo." *Oryx* 36(1): 35–40.

Durland, William. 1922. "The Forests of the Dominican Republic." *Geographical Review* 7(2): 206–222.

Eba'a Atyi, Richard, and M. Simula. 2002. "Forest Certification: Pending Challenges for Tropical Timber." ITTO International Workshop on Comparability and Equivalence of Forest Certification Schemes. Kuala Lumpur, Malaysia.

Echavarria, Fernando. 1998. "Monitoring Forests in the Andes Using Remote Sensing: An Example from Southern Ecuador." Pp. 100–120 in K. Zimmerer and K. Young (eds.). *Nature's Geography: New Lessons for Conservation in Developing Countries.* Madison, Wisc.: University of Wisconsin Press.

Edelman, Marc. 1995. "Rethinking the Hamburger Thesis: Deforestation and the Crisis of Central America's Beef Exports." Pp. 25–62 in M. Painter and Wm. Durham (eds.). *The Social Causes of Environmental Destruction in Latin America.* Ann Arbor, Mich.: The University of Michigan Press.

Eden, Michael J. 1990. *Ecology and Land Management in Amazonia.* London: Belhaven Press.

Ehui, Simeon, T. Hertel, and P. Preckel. 1990. "Forest Resource Depletion, Soil Dynamics, and Agricultural Productivity in the Tropics." *Journal of Environmental Economics and Management* 18: 136–154.

Ellis, Stephen. 1998. "Liberia's Warlord Insurgency." Pp. 155–171 in C.

Clapham (ed.). *African Guerrillas*. Bloomington, Ind.: Indiana University Press.

Erhardt-Martinez, Karen. 1998. "Social Determinants of Deforestation in Developing Countries: A Cross-National Study." *Social Forces* 77(2): 567–589.

Erhardt-Martinez, Karen, E. Crenshaw, and J. C. Jenkins. 2002. "Deforestation and the Environmental Kuznets Curve: A Cross-National Investigation of Intervening Mechanisms." *Social Science Quarterly* 83(1): 226–43.

Eyre, Lawrence A. 1987. "Jamaica: Test Case for Tropical Deforestation." *Ambio* 16(6): 338–343.

Fa, John. 1991. *Conservacion de los Ecosistemas Forestales de Guinea Ecuatorial*. Gland, Switzerland: UICN (The World Conservation Union).

Fairhead, James, and M. Leach. 1996. *Misreading the African Landscape: Society and Ecology in a Forest-Savanna Mosaic*. Cambridge, England: Cambridge University Press.

———. 1998. *Reframing Deforestation: Global Analyses and Local Realities in West Africa*. London: Routledge.

Faminow, Merle. 1998. *Cattle, Deforestation, and Development in the Amazon: An Economic, Agronomic, and Environmental Perspective*. London: CAB International.

Faminow, Merle, and K. Klein. 2000. "Adoption of Agro-Forestry in Nagaland, India Using Farmer-Led Technology Development and Dissemination." *Canadian Journal of Agricultural Economics* 48(5): 585–595.

Fearnside, Philip. 2003. "Conservation Policy in Brazilian Amazon: Understanding the Dilemmas." *World Development* 31(5): 757–779.

Food and Agricultural Organization (FAO). 1948. *Forest Production Yearbook*. Rome.

———. 1982a. "Tropical Forest Resources." FAO Forestry Paper, #30. Rome.

———. 1982b. *Forest Resources Assessment Project—Forest Resources of Tropical Africa, Regional Synthesis*. Rome.

———. 1993. "Forest Resources Assessment, 1990—Tropical Countries." FAO Forestry Paper, #112. Rome.

———. 2001. "Forest Resources Assessment, 2000." FAO Forestry Paper #140. Rome.

———. 2003a. "Country Profiles—Gambia—Plantations." Available at www.fao.org/forestry/fo/country.

———. 2003b. "Forestry Outlook Study for Africa (FOSA)." Subregional Report: Central Africa. Rome.

Forsyth, Adrian, and K. Miyata. 1983. *Tropical Nature: Life and Death in the Rain Forests of Central and South America*. New York: Simon and Schuster.

Foster, Andrew D., and M. Rosenzweig. 2003. "Economic Growth and the Rise of the Forests." *Quarterly Journal of Economics* 118(May): 601–637.

Fox, Jefferson. 1993. "Forest Resources in a Nepali Village in 1980 and 1990:

The Positive Influence of Population Growth." *Mountain Research and Development* 13(1): 89–98.

———. 2001. "Misunderstanding Deforestation: How Blaming 'Slash and Burn' Farmers is Deforesting Mainland Southeast Asia." Presentation to the Ecology and Culture Seminar, Columbia University, New York, October.

Fox, Jefferson, J. Krummel, S. Yarnasarn, M. Ekasingh, and N. Podger. 1995. "Land Use and Landscape Dynamics in Northern Thailand: Assessing Change in Three Upland Watersheds." *Ambio* 24(6): 328–334.

Fox, Jefferson, S. Leisz, D. Truong, A. Rambo, N. Tuyen, and L. Cuc. 2001. "Shifting Cultivation Without Deforestation: A Case Study in the Mountains of Northwestern Vietnam." *Kluwer International Series in Engineering and Computer Science* 626: 289–308.

Fugisaka, Sam, W. Bell, N. Thomas, L. Hurtado, and E. Crawford. 1996. "Slash and Burn Agriculture, Conversion to Pasture, and Deforestation in Two Brazilian Amazon Colonies." *Agriculture, Ecosystems, and Environment* 59(1/2): 115–125.

Fundacao Nacional do Indio (FUNAI). 2001. "As Terras Indigenas." Available at http://www.funai.gov.br.

Gami, Norbert, and R. Nasi. 2001. "Sustainability and Security of Intergenerational Access to Resources: Participatory Mapping Studies in Gabon." Pp. 214–228 in Carol Pierce Colfer and Yvonne Byron (eds.). *People Managing Forests*. Washington, D.C.: Resources for the Future.

Garcia, Jelson, and L. Mulkins. 1997. "Ecological Management of a Residual Forest in Negros Occidental, Philippines: The Experience of Marco Flores." Accessed at www.reap-canada.com/Reports/macking.doc.

Geddes, William Robert. 1976. *Migrants of the Mountains: The Cultural Ecology of the Blue Miao (Hmong Njua) of Thailand*. Oxford, England: Clarendon Press.

Geist, Helmut J., and E. Lambin. 2001. "What Drives Tropical Deforestation? A Meta-Analysis of Proximate and Underlying Causes of Deforestation based on Subnational Case Study Evidence." Land Use and Cover Change (LUCC) Report #4, Louvain-la-Neuve, Belgium.

———. 2002. "Proximate Causes and Underlying Driving Forces of Tropical Deforestation." *Bioscience* 52(2): 143–150.

Gibson, Clark C., and C. Becker. 2000. "A Lack of Institutional Demand: Why a Strong Local Community in Western Ecuador Fails to Protect Its Forest." Pp. 135–161 in C. Gibson, M. McKean, and E. Ostrom (eds.). *People and Forests: Communities, Institutions, and Governance*. Cambridge, Mass.: MIT Press.

Gibson, Clark C., M. McKean, and E. Ostrom. 2000. *People and Forests: Communities, Institutions, and Governance*. Cambridge, Mass.: MIT Press.

Giddens, Anthony. 1984. *The Constitution of Society: Outline of the Theory of Structuration*. Berkeley, Calif.: University of California Press.

Gober, Patricia. 2000. "In Search of Synthesis." *Annals of the Association of American Geographers* 90(1): 1–11.

Graybeal, Frederick T. 2001. "Evolution of Environmental Practice During Exploitation at the Camp Caiman Gold Project in French Guiana." Pp. 222–232 in I. Bowles and G. Prickett (eds.). *Footprints in the Jungle: Natural Resource Industries, Infrastructure, and Biodiversity Conservation.* New York: Oxford University Press.

Grinker, Roy R. 1994. *Houses in the Rain Forest: Ethnicity and Inequality Among Farmers and Foragers in Central Africa.* Berkeley, Calif.: University of California Press.

Grossman, Lawrence S. 1998. *The Political Ecology of Bananas: Contract Farming, Peasants, and Agrarian Change in the Eastern Caribbean.* Chapel Hill: University of North Carolina Press.

Guha, Ramachandra. 2000. *The Unquiet Woods: Ecological Change and Peasant Resistance in the Himalaya.* Expanded edition. Berkeley: University of California Press.

Guha, Ramachandra, S. Prasad, and M. Gadgil. 1984. "Deforestation and Degradation of Natural Plant Resources in India." *Journal of Indian Anthropology and Sociology* 19: 246–253.

Hairiah, Kurniaturn. 2000. "Shade Based Control of *Imperata cylindrical* by *Mucuna pruriens.*" Cornell International Institute for Food, Agriculture, and Development (CIFAD), Annual Report, 1999–2000.

Harcourt, Caroline, and Jeffrey Sayer (eds.). 1996. *The Conservation Atlas of Tropical Forests: The Americas.* New York: Simon and Schuster.

Hardiman, David. 1996. "Farming in the Forest: The Dangs 1830–1992." Pp. 101–131 in M. Poffenerger and B. McGean (eds.). *Village Voices, Forest Choices: Joint Forest Management in India.* Delhi: Oxford University Press.

Harms, Robert. 1981. *River of Wealth, River of Sorrow: The Central Zaire Basin in the Era of the Slave and Ivory Trade.* New Haven: Yale University Press.

Hecht, Susanna. 2002. "Solutions or Drivers? The Dynamics and Implications of Bolivian Lowland Deforestation." Unpublished paper.

———. 2004. "Invisible Forests: The Political Ecology of Forest Resurgence in El Salvador." Pp. 64–103 in R. Peet and M. Watts (eds.). *Liberation Ecologies: Environment, Development, and Social Movements,* second edition. London: Routledge.

Hecht, Susanna, A. Anderson, and P. May. 1988. "The Subsidy from Nature: Shifting Cultivation, Successional Palm Forests, and Rural Development." *Human Organization* 47(1): 25–35.

Hecht, Susanna, and A. Cockburn. 1989. *The Fate of the Forest: Developers, Destroyers, and Defenders of the Amazon.* London: Verso.

Hegen, E. E. 1966. *Highways into the Upper Amazon Basin: Pioneer Lands in Southern Colombia, Ecuador, and Northern Peru.* Gainesville, Fla.: University Press of Florida.

Hill, Polly. 1963. *The Migrant Cocoa Farmers of Southern Ghana.* Cambridge, England: Cambridge University Press.

Hiroaka, Mario. 1995. "Land Use in the Amazon Estuary." *Global Environmental Change* 5(4): 323–336.

Hiroaka, Mario, and S. Yamamoto. 1980. "Agricultural Development in the Upper Amazon of Ecuador." *Geographical Review* 70(4): 423–446.

Holden, Stein. 1997. "Adjustment Policies, Peasant Household Resource Allocation, and Deforestation in Northern Zambia: An Overview and Some Policy Conclusions." *Forum for Development Studies* 1: 117–134.

Holden, Stein, and Helga Huoslef. 1995. "Transmigration Settlements in Seberida: Causes and Consequences of the Deterioration of Farming Systems of Settlers in a Rain Forest Environment." Pp. 107–125 in O. Sandbukt (ed.). *Management of Tropical Forests: Towards an Integrated Perspective*. University of Oslo, Oslo, Norway: Centre for Development and the Environment.

Holmgren, Peter, E. Masakha, and H. Sjoholm. 1994. "Not All African Land Is Being Degraded: A Recent Survey of Trees on Farms in Kenya Reveals Rapidly Increasing Forest Resources." *Ambio* 23(7): 390–395.

Humphries, Sally. 1998. "Milk Cows, Migrants, and Land Markets: Unraveling the Complexities of Forest to Pasture Conversion in Northern Honduras." *Economic Development and Cultural Change* 47(1): 95–124.

Hussain, Talib, P. Bibi, and P. Kaushal. 1999. "We Are All Part of the Same 'Kudrat': Community Forest Management in Rajaji National Park." *Forests, Trees, and People* 38: 35–38.

Hyde, W. F., G. Amacher, and W. Magrath. 1996. "Deforestation and Forest Land Use: Theory, Evidence, and Policy Implications." *World Bank Research Observer* 11(2): 223–248.

Instituto Brasileiro de Geografia y Estatistica (IGBE). 2001. "Censo Agropecuario de 1995–1996." Brasilia. Available at http://www1.ibge.gov.br.

Instituto Nacional de Pesquisas Espaciales. 2004. "Monitoramento da Floresta Amazonica Brasileira por Satelite: Projeto Prodes." Accessed at www.obt.inpe.br/prodes.

Jackson, W. J., R. Tamrakar, S. Hunt, and K. Shepherd. 1998. "Land Use Changes in Two Middle Hill Districts of Nepal." *Mountain Research and Development* 18(3): 193–212.

Jacobs, Marius. 1988. *The Tropical Rain Forest: A First Encounter*. Berlin: Springer-Verlag.

Jaeger, Carlos, O. Renn, E. Rosa, and T. Webler. 2001. *Risk, Uncertainty, and Rational Action*. London: Earthscan.

Jaramillo Alvarado, P. 1936. *Tierras del Oriente: Caminos, Ferrocarriles, Adminstracion, Riqueza Aurifera*. Quito, Ecuador: Imprenta y Encuadernacion Nacionales.

Jepson, Paul, James K. Jarvie, Kathy MacKinnon, and Kathryn Monk. 2001. "The End for Indonesa's Lowland Forests?" *Science* 292: 859–861.

Jones, Jeffrey. 1989. "Human Settlement of Tropical Colonization in Central America." Pp. 43–85 in Debra A. Schumann and Wm. L. Partridge (eds.). *The Human Ecology of Tropical Land Settlement in Latin America*. Boulder, Colo.: Westview Press.

Jones, P. J., A. Burlison, and A. Tye. 1991. *Conservacao dos Ecosistemas Florestais na Republica Democratica de Sao Tome e Principe*. Gland, Switzerland: UICN (The World Conservation Union).

Jones, D. W., and R. O'Neill. 1994. "Development Policies, Rural Land Use, and Tropical Deforestation." *Regional Science and Urban Economics* 24: 753–771.

Kahn, James, and J. McDonald. 1995. "Third World Debt and Tropical Deforestation." *Ecological Economics* 12: 107–123.

Kaimowitz, David. 1996. *Livestock and Deforestation: Central America in the 1980s and 1990s: A Policy Perspective*. Jakarta: Center for International Forestry Research (CIFOR).

———. 2002. "The Meaning of Johannesburg." Center for International Forestry Research (CIFOR)–Polex Listserve, September 26th. Accessed at www.cifor.cgiar.org/docs/polex.

Kaimowitz, David, and A. Angelsen. 1998. *Economic Models of Tropical Deforestation: A Review*. Bogor, Indonesia: Center for International Forestry Research (CIFOR).

Kaimowitz, David, A. Faune, and R. Mendoza. 1999. Who Owns the Forest? Rulers and Resources in Northeast Nicaragua. Unpublished manuscript, Center for International Forestry Research (CIFOR).

Kaimowitz, David, and J. Smith. 2001. "Soybean Technology and the Loss of Natural Vegetation in Brazil and Bolivia." Pp. 195–212 in A. Angelsen and D. Kaimowitz (eds.). *Agricultural Technologies and Tropical Deforestation*. London: CAB International.

Kaimowitz, David, C. Vallejos, P. Pacheco, and R. Lopez. 1999. "Municipal Governments and Forest Management in Lowland Bolivia." *Journal of Environment and Development* 7(1): 45–59.

Kampayana, Theobald. 1992. "Rarete des Terres et Strategies d'Utilisation du Bois: Resultats d'une Enquete Pontuelle dans les Menages Agricoles Ruraux au Rwanda." Kigali, Rwanda: Ministere de l'Agriculture et de l'Elevage.

Kanel, K. R., and K. Shrestha. 2001. "Tropical Secondary Forests in Nepal and Their Importance to Local People." *Journal of Tropical Forest Science* 13(4): 691–704.

Karanth, K. Ullas. 2002. "Nagarahole: Limits and Opportunities in Wildlife Conservation." Pp. 189–202 in John Terborgh, Carel van Schaik, Lisa Davenport, and Madhu Rao (eds.). *Making Parks Work: Strategies for Preserving Tropical Nature*. Washington, D.C.: Island Press.

Katzer, Jeffrey, K. Cook, and W. Crouch. 1998. *Evaluating Information: A Guide for Users of Social Science Research*. Boston, Mass.: McGraw-Hill.

Kebron, T., and L. Hedlund. 2000. "Land Cover Changes Between 1958 and 1986 in Kalu District, Southern Wello, Ethiopia." *Mountain Research and Development* 20(1): 42–51.

Kepner, Charles D., and J. Soothill. 1935 (reissued 1967). *The Banana Empire: A Case Study of Economic Imperialism*. New York: Russell and Russell.

Kingsley, Mary. 1897. *Travels in West Africa: Congo Francais, Corisco, and Cameroons*. London: MacMillan.

Klooster, Daniel. 2000. "Beyond Deforestation: The Social Context of Forest Change in Two Indigenous Communities in Highland Mexico." *Yearbook, Conference of Latin Americanist Geographers* 26: 47–59.

Kohlin, Gunnar, and Peter J. Parks. 2001. "Spatial Variability and Disincentives to Harvest: Deforestation and Fuelwood Collection in South Asia." *Land Economics* 77(2): 206–218.

Kotto-Same, J, A. Moukam, R. Njomgang, T. Tiki-Manga, J. Toonye, C. Diaw, C. Gockowski, S. Hauser, S. Weise, D. Nwaga, L. Zapfack, C. Palm, P. Woomer, A. Gillison, D. Bignell, and J. Tondoh. 2000. *Alternatives to Slash and Burn: Summary Report and Synthesis of Phase II in Cameroon.* Nairobi: International Centre for Research in Agroforestry (ICRAF).

Kramer, R., K. Van Schaik, and J. Johnson. 1997. *Last Stand: Protected Areas and the Defense of Tropical Biodiversity.* New York: Oxford University Press.

Krutilla, K., Wm. Hyde, and D. Barnes. 1995. "Peri-Urban Deforestation in Developing Countries." *Forest Ecology and Management* 74: 181–195.

Kummer, David. 1992. "Upland Agriculture, the Land Frontier, and Forest Decline in the Philippines." *Agroforestry Systems* 18: 31–46.

Kummer, David, R. Concepcion, and B. Canizares. 1994. "Environmental Degradation in the Uplands of Cebu." *Geographical Review* 84(3): 266–276.

Kunstadter, Peter, and E. C. Chapman. 1978. "Problems of Economic Development and Shifting Cultivation in Northern Thailand." Pp. 3–23 in Peter Kunstadter, E. C. Chapman, and Sanga Sabhasri (eds.). *Farmers in the Forest: Economic Development and Marginal Agriculture in Northern Thailand.* Honolulu: University of Hawaii Press.

Lambin, Eric, B. Turner, H. Geist, S. Agbola, A. Angelsen, J. Bruce, O. Coomes, R. Dirzo, G. Fischer, C. Folke, P. George, K. Homewood, J. Imbernon, R. Leemans, X. Li, E. Moran, M. Mortimore, P. Ramakrishnan, J. Richards, H. Skanes, W. Steffen, G. Stone, U. Svedin, T. Veldkamp, C. Vogel, and J. Xu. 2001. "The Causes of Land-Use and Land-Cover Change: Moving Beyond the Myths." *Global Environmental Change* 11: 261–269.

Larson, Anne. 2000. "Peasant Agroforesters: Fiction or Reality?" Paper presented at the meetings of the Latin American Studies Association, Miami, March.

Lathrap, Donald. 1970. *The Upper Amazon.* London: Thames and Hudson.

Laurence, William F., M. Cochrane, S. Bergen, P. Fearnside, P. Delamonica, C. Barber, S. D'Angelo, and T. Fernandes. 2001. "The Future of the Brazilian Amazon." *Science* 291: 438–439.

Leach, Melissa. 1994. *Rainforest Relations: Gender and Resource Use Among the Mende of Gola, Sierra Leone.* Washington, D.C.: Smithsonian Institution Press.

Lee, Liu. 1999. "Labor Location, Conservation, and Land Quality: The Case of West Jilin, China." *Annals of the Association of American Geographers* 89(4): 633–658.

Likaka, Osumaka. 1997. *Rural Society and Cotton in Colonial Zaire.* Madison, Wisc.: University of Wisconsin Press.

Little, Daniel. 2000. "Explaining Large-Scale Historical Change." *Philosophy of the Social Sciences* 30(1): 89–112.

Liu, Dawning S., L. Iverson, and S. Brown. 1993. "Rates and Patterns of Deforestation in the Philippines: An Application of Geographic Information Analysis." *Forest Ecology and Management* 57(1): 1–16.

Livingstone, David. 1970 (1874). *The Last Journals of David Livingstone*. 2 vols. Westport, Conn.: Greenwood Press.

Lugo, Ariel. 1992. "Tree Plantations for Rehabilitating Damaged Forest Lands in the Tropics." Pp. 247–255 in Mohan Wali (ed.). *Ecosystem Rehabilitation, Volume 2: Ecosystem Analysis and Synthesis*. The Hague, Netherlands: SPB Academic.

Madsen, Berit. 2003. "The Evening School Teacher and His Dream for His New Born Child." Archived at www.msnepal.org/stories_articles/dalits.

Mahar, Dennis J., and C. Ducrot. 1998. "Land Use Zoning on Tropical Frontiers: Emerging Lessons from the Brazilian Amazon." Economic Development Institute Case Study, World Bank, Washington, D.C.

Maki, Sanna, R. Kalliola, and K. Vuorinen. 2001. "Road Construction in the Peruvian Amazon: Process, Causes, and Consequences." *Environmental Conservation* 28(3): 199–214.

Malaisse, F., and K. Binzangi. 1985. "Wood as a Source of Fuel in Upper Shaba (Zaire)." *Commonwealth Forestry Review* 64(3): 227–239.

Martin, Claude. 1991. *The Rain Forests of West Africa: Ecology, Threats, and Conservation*. Basel: Birkhauser Verlag.

Marx, Karl. 1963 (1852). *The 18th Brumaire of Louis Bonaparte*. New York: International Publishers.

Mather, A., and C. Needle. 1998. "The Forest Transition: A Theoretical Basis." *Area* 30(2): 117–124.

Mather, A., C. Needle, and J Fairbairn. 1999. "Environmental Kuznets Curves and Forest Trends." *Geography* 84(1): 55–65.

Mathews, Emily. 2001. "Understanding the FRA 2000, Forest Briefing No. 1." Washington, D.C.: World Resources Institute.

Maturana, Julia. 1999. "Mercado de Tierras en Brasil: Caso San Felix de Xingu." Occasional Paper, Center for International Forestry Research (CIFOR), Bogor, Indonesia.

McCann, James C. 1999. *Green Land, Brown Land, Black Land: An Environmental History of Africa, 1800–1990*. Portsmouth, N.H.: Heinemann.

McCarthy, J. F. 2000. "'Wild Logging': The Rise and Fall of Logging Networks and Biodiversity Conservation Projects on Sumatra's Rain Forest Frontier." Occasional Paper #31, Center for International Forestry Research (CIFOR), Bogor, Indonesia.

McConnell, William J., and Emilio Moran (eds.). 2000. *Meeting in the Middle: The Challenge of Meso-Level Integration. Land Use and Cover Change (LUCC)*. Report Series #5. Bloomington, Ind.: LUCC Focus 1 Office.

McPeak, J. G., and C. B. Barrett. 2001. "Differential Risk Exposure and Stochastic Poverty Traps Among East African pastoralists." *American Journal of Agricultural Economics* 83(3): 674–679.

McShane, Thomas O. 1999. "Voyages of Discovery: Four Lessons from the DGIS-WWF Tropical Forest Portfolio." *Arborvitae* (Supplement) Gland, Switzerland: World Wide Fund for Nature.

Medley, K. E. 1998. "Landscape Change and Resource Conservation Along the Tana River, Kenya." Pp. 39–55 in K. Zimmerer and T. Young (eds.). *Nature's Geography: New Lessons for Conservation in Developing Countries*. Madison, Wisc.: University of Wisconsin Press.

Meggers, Betty. 1971. *Amazonia: Man and Culture in a Counterfeit Paradise*. Arlington, Heights, Ill.: AHM Publishing.

Mertens, Benoit, and E. Lambin. 2000. "Land-Cover Change Trajectories in Southern Cameroon." *Annals of the Association of American Geographers* 90(3): 467–494.

Mertens, Benoit, R. Poccard-Chapuis, M. G. Piketty, A. Lacques, and A.Venturieri. 2002. "Crossing Spatial Analyses and Livestock Economics to Understand Deforestation Processes in the Brazilian Amazon: The Case of Sao Felix do Xingu in South Para." *Agricultural Economics* 27: 269–294.

Merton, Robert K. 1968. *Social Theory and Social Structure*. New York: Free Press.

Meyer, W. B., and B. L. Turner, II. 1992. "Human Population Growth and Global Land Use/Land Cover Change." *Annual Review of Ecology and Systematics* 23: 39–61.

Migdal, Joel. 1988. *Strong Societies and Weak States*. Princeton, N.J.: Princeton University Press.

Mill, John Stuart. 1865. *A System of Logic*. London: Longmans.

Mintz, Sidney. 1986. *Sweetness and Power: The Place of Sugar in Modern History*. New York: Penguin Books.

Misana, Salome, C. Mung'ong'o, and B. Mukamuri. 1996. "Miombo Woodlands in the Wider Context: Macro-Economic and Inter-Sectoral Influences." Pp. 73–99 in Bruce Campbell (ed.). *The Miombo in Transition: Woodlands and Welfare in Africa*. Bogor, Indonesia: Center for International Forestry Research (CIFOR).

Mittermeier, Russell, N. Myers, P. Robles Gil, and C. Mittermeier. 2000. *Hotspots: Earth's Biologically Richest and Most Endangered Terrestrial Ecoregions*. Mexico City: Cemex.

Molotch, Harvey, W. Freudenburg, and K. Paulsen. 2000. "History Repeats Itself, but How? City Character, Urban Tradition, and the Accomplishment of Place." *American Sociological Review* 65(6): 791–823.

Monaghan, Paul F. 2000. "Peasants, the State, and Colonization of Haiti's Last Rainforest." Paper presented at the meetings of the Latin American Studies Association, Miami.

Monela, G. C., G. Kajembe, A. Kaoneka, and G. Kowero. 1999. "Household Livelihood Strategies in the Miombo Woodlands of Tanzania: Emerging Trends." *Tanzanian Journal of Forestry and Nature Conservation* 73: 17–33.

Moore, Henrietta, and M. Vaughan. 1994. *Cutting Down Trees: Gender, Nutrition, and Agricultural Change in the Northern Province of Zambia, 1890–1990.* Portsmouth, N.H.: Heinemann.

Moran, Emilio, E. Brondizio, and S. McCracken. 2002. "Trajectories of Land Use: Soils, Succession, and Crop Choice." Pp. 193–217 in C. Wood and R. Porro (eds.). *Deforestation and Land Use in the Amazon.* Gainesville, Fla.: University Press of Florida.

Mukherjee, S. D. 1997. "Is Handing Over Forests to Local Communities a Solution to Deforestation? Experience in Andhra Pradesh–India." *Indian Forester* 123: 460–471.

Müller, D., and M. Zeller. 2002. "Land Use Dynamics in the Central Highlands of Vietnam: A Spatial Model Combining Village Survey Data with Satellite Imagery Interpretation." *Agricultural Economics* 27: 333–354.

Myers, Norman. 1979. *Conversion of Tropical Moist Forests.* Washington, D.C.: National Academy Press.

———. 1981. "The Hamburger Connection: How Central America's Forests Become North America's Hamburgers." *Ambio* 10(1): 3–8.

———. 1984. *The Primary Source: Tropical Forests and Our Future.* New York: Norton.

Nepstad, Daniel, G. Carvalho, A Barras, A. Alencar, J. Capobianco, J. Bishop, P. Moutinho, D. Lefebre, U. Silva, and E. Prins. 2001. "Road Paving, Fire Regime Feedbacks, and the Future of Amazon Forests." *Forest Ecology and Management* 154(3): 395–407.

Nepstad, Daniel, D. McGrath, A. Alencar, A. Barros, G. Carvalho, M. Santilli, and M. del C. Vera Diaz. 2002. "Frontier Governance in Amazonia." *Science* 295: 629–631.

Nepstad, Daniel, C. Uhl, and E. Serrao. 1991. "Recuperation of a Degraded Amazon Landscape: Forestry Recovery and Agricultural Restoration." *Ambio* 20: 248–255.

Nepstad, Daniel, C. Verissimo, A. Alencar, A. Nobres, C. Lima, E. Lefebvre, P. Schlesinger, P. Potter, C. Moutinho, and P. Mendoza. 1999. "Large-Scale Impoverishment of Amazonian Forests by Logging and Fire." *Nature* 398: 505–508.

Netting, R. Mac. 1993. *Smallholders, Householders: The Ecology of Small Scale, Sustainable Agriculture.* Stanford, Calif.: Stanford University Press.

Ng'weno, Bettina. 2001. "Reidentifying Ground Rules: Community Inheritance Disputes Among the Digo of Kenya." Pp. 111–137 in A. Agrawal and C. Gibson (eds.). *Communities and the Environment: Ethnicity, Gender, and the State in Community Based Conservation.* New Brunswick, N.J.: Rutgers University Press.

Nibbering, J. W. 1999. "Tree Planting on Deforested Farmland, Sewu Hills, Java, Indonesia: Impact of Economic and Institutional Changes." *Agroforestry Systems* 46: 65–82.

North, Douglass. 1990. *Institutions, Institutional Change, and Economic Performance.* Cambridge, England: Cambridge University Press.

Nyerges, Endre A., and G. Green. 2000. "The Ethnography of Landscape: GIS and Remote Sensing in the Study of Forest Change in West African Guinea Savanna." *American Anthropologist* 102(2): 271–289.

Oates, John F. 1999. *Myth and Reality in the Rain Forest: How Conservation Strategies Are Failing in West Africa.* Berkeley, Calif.: University of California Press.

O'Brien, Karen L. 1998. *Sacrificing the Forest: Environmental and Social Struggles in Chiapas.* Boulder, Colo.: Westview Press.

Okoth-Ogendo, H. W. 1986. "Tenure Issues in Spontaneous Settlement." Pp. 170–75 in *Spontaneous Settlement Formation,* Vol. II: *Case Studies.* Nairobi: United Nations.

Ortiz, Sutti. 1984. "Colonization in the Colombian Amazon." Pp. 204–230 in M. Schmink and C. Woods (eds.). *Frontier Expansion in Amazonia.* Gainesville, Fla.: University Press of Florida.

Ortiz-Chour, Hivy. 1999. "Nepal." Working Paper #12. Rome: Forest Resources Assessment Programme, Food and Agricultural Organization of the United Nations.

Ostrom, Elinor. 1999. "Self-Governance and Forest Resources." Occasional Paper #20. Bogor, Indonesia: Center for International Forestry Research (CIFOR).

Owusu, J. Henry. 1998. "Current Convenience, Desperate Deforestation: Ghana's Adjustment Program and the Forestry Sector." *Professional Geographer* 50(4): 418–436.

Padoch, Christine, and K. Coffey. 2003. "Monitoring the Demise of Swidden in Southeast Asia." Pp. 103–124 in *Local Land Use Strategies in a Globalizing World: Shaping Sustainable Social and Natural Worlds,* Vol. I, *Proceedings of the International Conference,* August 21–23, 2003. Copenhagen, Denmark: Institute of Geography.

Paige, Jeffrey. 1997. *Coffee and Power: Revolution and the Rise of Democracy in Central America.* Cambridge, Mass.: Harvard University Press.

Parker, Dawn, T. Berger, and S. Manson (eds.). 2002. "Agent-Based Models of Land-Use and Land-Cover Change." LUCC Report Series No. 6. Louvain-la-Neuve, Belgium.

Parker, Dawn, S. Manson, M. Jansesen, M. Hoffmann, and P. Deadman. 2003. "Multi-Agent Systems for the Simulation of Land-Use and Land-Cover Change: A Review." *Annals of the Association of American Geographers* 93(2): 314–337.

Parkin, David J. 1972. *Palms, Wine, and Witnesses: Public Spirit and Private Gain in an African Farming Community.* San Francisco, Calif.: Chandler.

Parks, Peter, E. Barbier, and J. Burgess. 1998. "The Economics of Forest Land Use in Temperate and Tropical Areas." *Environmental and Resource Economics* 11(3–4): 473–487.

Pearce, Fred. 1998. "Beyond Hope: Poverty and Politics Are Putting an End to Rainforest Conservation." *New Scientist* October 31, 1998.

Peluso, Nancy. 1992. *Rich Forests, Poor People: Resource Control and Resistance in Java*. Berkeley: University of California Press.

Pelzer, Karl. 1945. *Pioneer Settlement in the Asiatic Tropics: Studies in Land Utilization and Agricultural Colonization in Southeast Asia*. New York: American Geographical Society.

———. 1968. "Man's Role in Changing the Landscape of Southeast Asia." *Journal of Asian Studies* 27(2): 269–279.

Persson, Reidar. 1996. "Tropical Plantations: Success or Failure?" *IDRC Currents* Ottawa: International Development Research Center.

Perz, S. G. 2001. "Household Demographic Factors as Life Cycle Determinants of Land Use in the Amazon." *Population Research and Policy Review* 20(3): 159–186.

———. 2002. "Population Growth and Net Migration in the Brazilian Legal Amazonia, 1970–1996." Pp. 107–132 in C. Wood and R. Porro (eds.) *Deforestation and Land Use in the Amazon*. Gainesville, Fla.: University Press of Florida.

Peterson, Richard. 2000. *Conversations in the Rainforest: Culture, Values, and the Environment in Central Africa*. Boulder, Colo.: Westview Press.

Pfaff, A., S. Kerr, R. Hughes, S. Liu, A. Sanchez, D. Schimel, J. Tosi, and V. Watson. 2000. "The Kyoto Protocol and Payments for Tropical Forest: An Interdisciplinary Method for Estimating Carbon-Offset Supply and Increasing the Feasibility of a Carbon Market Under CDM." *Ecological Economics* 35(2): 203–221.

Pichon, Francisco, C. Marquette, L. Murphy, and R. Bilsborrow. 2001. "Land Use, Agricultural Technology, and Deforestation Among Settlers in the Ecuadorian Amazon." Pp. 153–166 in A. Angelsen and D. Kaimowitz (eds.). *Agricultural Technologies and Tropical Deforestation*. New York: CAB International.

Pimm, Stuart, M. Ayres, A. Balmford, G. Branch, K. Brandon, T. Brooks, R. Bustamante, R. Costanza, R. Cowling, L. Curran, A. Dobson, S. Farber, G. da Fonseca, C. Gascon, R. Kitching, J. McNeely, T. Lovejoy, R. Mittemeier, N. Myers, J. Patz, B. Raffle, D. Rapport, P. Raven, C. Roberts, J. Rodriguez, A. Rylands, C. Tucker, C. Safina, C. Samper, M. Stiassny, J. Supriatna, D. Wall, and D. Wilcove. 2001. "Can We Defy Nature's End?" *Science* 293: 2207.

Pinedo-Vasquez, Miguel, D. Zarin, K. Coffey, C. Padoch, and F. Rabelo. 2001. "Post-Boom Logging in Amazonia." *Human Ecology* 29(2): 219–239.

Plumwood, Val, and R. Routley. 1982. "World Rainforest Destruction: The Social Factors." *The Ecologist* 12(1): 4,22.

Poffenberger, Mark, and Ajit Banerjee. 1996. "Conclusion." Pp. 324–332 in M. Poffenberger and B. McGean (eds.). *Village Voices, Forest Choices: Joint Forest Management in India*. Delhi: Oxford University Press.

Poffenberger, Mark, and B. McGean (eds.). 1996. *Village Voices, Forest Choices: Joint Forest Management in India*. Delhi: Oxford University Press.

Poffenberger, Mark, B. McGean, and A. Khare. 1996. "Communities Sustaining India's Forests in the Twenty-first Century." Pp. 17–55 in M. Poffenberger and B. McGean (eds.). *Village Voices, Forest Choices: Joint Forest Management in India*. Delhi: Oxford University Press.

Poffenberger, Mark, N. Ravindrath, D. Pandey, I. Murthy, R. Bist, and D. Jain. 2002. "Communities and Climate Change: The Clean Development Mechanism and Village-Based Forest Restoration in Central India." Accessed at www.cifor.cgiar.org/docs/_ref/polex.

Poffenberger, Mark, and C. Singh. 1996. "Communities and the State: Re-Establishing the Balance in Indian Forest Policy." Pp. 56–85 in M. Poffenberger and B. McGean (eds.). *Village Voices, Forest Choices: Joint Forest Management in India*. Delhi: Oxford University Press.

Poharel, Ridish, S. Adhikari, and Y. Thapa. 1999. "Sankarnagar Forest User Group: Learning from a Successful FUG in the Terai." *Forests, Trees, and People* 38: 25–32.

Polanyi, Karl. 1971 (1944). *The Great Transformation*. Boston: Beacon Press.

Poteete, Amy, and Elinor Ostrom. 2002. "An Institutional Approach to the Study of Forest Resources." International Forestry Resources and Institutions Research Program, Indiana University. Accessed at www.indiana.edu/~workshop/papers/W01.8 pdf.

Potter, Lesley, and Justin Lee. 1998. "Tree Planting in Indonesia: Trends, Impacts, and Directions." Occasional Paper #18, Center for International Forestry Research (CIFOR). Bogor, Indonesia.

Preston, David. 1998. "Post-Peasant Capitalist Graziers: The 21st Century in Southern Bolivia." *Mountain Research and Development* 18: 151–158.

Ragin, Charles. 1987. *The Comparative Method: Moving Beyond Qualitative and Quantitative Strategies*. Berkeley, Calif.: University of California Press.

———. 2000. *Fuzzy-Set Social Science*. Chicago: University of Chicago Press.

Raison, Jean-Paul. 1968. "La Colonizacion des Terres Nueves Intertropicales." *Etudes Rurales* 31: 5–112.

Rathore, S., S. Singh, J. Singh, and A. Tiwari. 1997. "Changes in Forest Cover in a Central Himalayan Catchment: Inadequacy of Assessment Based on Forest Area Alone." *Journal of Environmental Management* 49: 265–276.

Ravindranath, N. H., and D. Hall. 1994. "Indian Forest Conservation and Tropical Deforestation." *Ambio* 23(8): 521–523.

Rawat, A. 1995. "Deforestation and Forest Policy in the Lesser Himalayan Kumaun: Impacts on Peasant Women and Tribal Populations." *Mountain Research and Development* 15(4): 311–322.

Richards, John F. 1987. "Environmental Changes in Dehra Dun Valley, India: 1880–1980." *Mountain Research and Development* 7(3): 299–304.

Rocheleau, Dianne, L. Ross, J. Morrobel, and R. Hernandez. 1997. "Forests, Gardens, and Tree Farms: Gender, Class, and Community at Work in the Landscapes of Zambrana-Chacuey." Working Paper. Ecology, Gender, and Community Project. Clark University, Worcester, Mass.

Rohter, Larry. 2002. "Amazon Forests Still Burning Despite the Good Intentions." *New York Times*, August 23rd.

Rosero-Bixby, Luis, and A. Palloni. 1998. "Population and Deforestation in Costa Rica." *Population and Environment* 20(2): 149–184.

Rudel, Thomas K. 1995. "When Do Property Rights Matter? Open Access, Informal Social Controls, and Deforestation in the Ecuadorian Amazon." *Human Organization* 54(2): 187–194.

———. 1998. "Is There a Forest Transition? Deforestation, Reforestation, and Development." *Rural Sociology* 63(4): 533–52.

———. 2002. "Paths of Destruction and Regeneration: Globalization and Forests in the Tropics." *Rural Sociology* 67(4): 622–636.

Rudel, Thomas K., Diane Bates, and Rafael Machinguiashi. 2002a. "A Tropical Forest Transition? Out-Migration, Agricultural Change, and Reforestation in the Ecuadorian Amazon." *Annals of the Association of American Geographers* 92(1): 87–102.

———. 2002b. "Ecologically Noble Amerindians: Cattle Ranching and Cash Cropping Among Shuar and Colonists in Ecuador." *Latin American Research Review* 37(1): 144–159.

Rudel, Thomas K., M. Perez-Lugo, and H. Zichal. 2000. "When Fields Revert to Forests: Development and Spontaneous Reforestation in Post-War Puerto Rico." *Professional Geographer* 52(3): 386–397.

Rudel, Thomas K., and J. Roper. 1996. "Regional Patterns and Historical Trends in Tropical Deforestation, 1976–1990: A Qualitative Comparative Analysis." *Ambio* 27(4): 160–166.

———. 1997a. "Paths to Rain Forest Destruction: Cross-National Patterns of Tropical Deforestation." *World Development* 25(1): 53–65.

———. 1997b. "Forest Fragmentation in the Humid Tropics." *Singapore Journal of Tropical Geography* 18(1): 99–109.

Rudel, Thomas K., with Bruce Horowitz. 1993. *Tropical Deforestation: Land Clearing and Small Farmers in the Ecuadorian Amazon.* New York: Columbia University Press.

Ruf, Francois. 2001. "Tree Crops as Deforestation and Reforestation Agents: The Case of Cocoa in Cote D'Ivoire and Sulawesi." Pp. 291–315 in A. Angelsen and D. Kaimowitz (eds.). *Agricultural Technologies and Tropical Deforestation.* London: CAB International.

Ruitenbeek, Jack, and Cynthia Cartier. 2001. "The Invisible Wand: Adaptive Co-management as an Emergent Strategy in Complex Bio-economic Systems." Occasional Paper #34, Center for International Forestry Research, Bogor, Indonesia.

Sader, Steven, and A. Joyce. 1988. "Deforestation Rates and Trends in Costa Rica, 1940–1983." *Biotropica* 20(1): 11–19.

Salam, M., T. Noguchi, and M. Koike. 2000. "Understanding Why Farmers Plant Trees in Homestead Agroforestry in Bangladesh." *Agroforestry Systems* 50(1): 77–93.

Santos-Granero, Fernando, and F. Barclay. 1998. *History, Economy, and Land Use in Peruvian Amazonia.* Washington, D.C.: Smithsonian Institution Press.

Sarin, Madhu. 2001. "Disempowerment in the Name of 'Participatory' Forestry? Village Forests Joint Management in Uttarakhand." *Forest, Trees, and People.* #44. Kontakt, Sweden: Swedish University of Agricultural Sciences.

Sayer, Jeffrey, C. Harcourt, and M. Collins (eds.). 1992. *The Conservation Atlas of Tropical Forests: Africa.* New York: Simon and Schuster.

Schnaiberg, Allan, and K. Gould. 1994. *Environment and Society: The Enduring Conflict.* New York: St. Martin's Press.

Schneider, Robert R., E. Arima, A. Verissimo, P. Barreto, and C. Souza, Jr. 2000. *Sustainable Amazon: Limitations and Opportunities for Rural Development.* Brasilia: World Bank and Imazon.

Schreier, Hans, S. Brown, M. Schmidt, P. Shah, B. Shrestha, G. Nakarmi, K. Subba, and S. Wymann. 1994. "Environmental Auditing: Gaining Forests but Losing Ground: A GIS Evaluation in a Himalayan Watershed." *Environmental Management* 18(1): 139–150.

Schroeder, Richard A. 1999. *Shady Practices: Agroforestry and Gender Politics in the Gambia.* Berkeley, California: University of California Press.

Schulze, Ernest, C. Wirth, and M. Heimann. 2000. "Managing Forests After Kyoto." *Science* 289: 2058–2059.

Schwartz, Norman B. 1987. "Colonization of Northern Guatemala: The Peten." *Journal of Anthropological Research* 43: 163–83.

Schweik, Charles, K. Adhikari, and K. Pandit. 1997. "Land-Cover Change and Forest Institutions: A Comparison of Two Sub-Basins in the Southern Siwalik Hills of Nepal." *Mountain Research and Development* 17(2): 99–116.

Scott, James C. 1975. *The Moral Economy of the Peasant.* New Haven, Conn.: Yale University Press.

———. 1985. *Weapons of the Weak: Everyday Forms of Peasant Resistance.* New Haven, Conn.: Yale University Press.

Sen, K. K., R. Semwal, U. Rana, S. Nautiyal, R. Maikhuri, K. Rao, and K. Saxena. 2002. "Patterns and Implications of Land Use/Cover Change: A Case Study of Pranmati Watershed (Garhwal Himalaya, India)." *Mountain Research and Development* 22(1): 56–62.

Shively, Gerald, and E. Martinez. 2001. "Deforestation, Irrigation, Employment, and Cautious Optimism in Southern Palawan, the Philippines." Pp. 335–346 in A. Angelsen and D. Kaimowitz (eds.). *Agricultural Technologies and Tropical Deforestation.* New York: CAB International.

Sierra, Rodrigo. 2000. "Dynamics and Patterns of Deforestation in the Western Amazon: The Napo Deforestation Front, 1986–1996." *Applied Geography* 20(1): 1–16.

Sierra, Rodrigo, and J. Stallings. 1998. "The Dynamics and Social Organization of Tropical Deforestation in Northwest Ecuador, 1983–1995." *Human Ecology* 26(1): 135–161.

Sikor, T., and D. Truong. 2002. "Agricultural Policy and Land Use Changes in a Black Thai Commune of Northern Vietnam, 1952–1997." *Mountain Research and Development* 22(3): 248–255.

Simkins, Paul D., and Frederick L. Wernstedt. 1971. "Philippine Migration: The Settlement of the Digos–Padada Valley, Davao Province." Monograph Series #16. Yale University Southeast Asian Studies Program, New Haven, Conn.

Singh, Neara M. 2002. "Federations of Community Forest Management Groups in Orissa: Crafting New Institutions to Assert Local Rights." Forests, Trees, and People #46. Kontakt: Swedish University of Agricultural Sciences.

Sivaramakrishnan, K. 1999. Modern Forests: Statemaking and Environmental Change in Colonial Eastern India. Stanford, Calif.: Stanford University Press.

Skole, David. 1999. "Does Place Matter? Deforestation Across Places." Paper Presented at the annual meeting of the Association of American Geographers, New York.

Sonnenfeld, David, and A. Mol (eds.). 2000. Ecological Modernization Around the World: Perspectives and Critical Debates. Portland, Ore., and London: Frank Cass.

Soussan, J., B. Shrestha, and L. Uprety. 1995. The Social Dynamics of Deforestation: A Case Study from Nepal. New York: Parthenon.

Southgate, Douglas. 1998. Tropical Forest Conservation: An Economic Assessment of the Alternatives in Latin America. New York: Oxford University Press.

Southworth, Jane, and C. Tucker. 2001. "The Influence of Accessibility, Local Institutions, and Socio-Economic Factors on Forest Cover Change in the Mountains of Western Honduras." Mountain Research and Development 21(3): 276–283.

Spruce, Richard. 1908. Notes of a Botanist on the Andes and the Amazon. 2 vols. London: MacMillan.

Stanley, R, and A. Neame (eds.). 1961. The Exploration Diaries of H. M. Stanley. New York: Vanguard Press.

Stokstad, Eric. 2001. "U.N. Report Suggests Slowed Forest Losses." Science 291: 2294.

Stolle, Fred, and Thomas P. Tomich. 1999. "The 1997–1998 Fire Event in Indonesia." Nature and Resources 35(3): 22–30.

Stone, Roger, and Claudia D'Andrea. 2001. Tropical Forests and the Human Spirit: Journeys to the Brink of Hope. Berkeley: University of California Press.

Struhsaker, Thomas. 1997. Ecology of an African Rain Forest: Logging in Kibale and the Conflict Between Conservation and Exploitation. Gainesville, Fla.: University Press of Florida.

Sunderlin, William. 1999. "The Effects of Economic Crisis and Political Change on Indonesia's Forest Sector, 1997–99." Occasional Paper. Center for International Forestry Research (CIFOR), Bogor, Indonesia.

Sunderlin, William, and I. Resosudarmo. 1999. "The Effect of Population and Migration on Forest Cover in Indonesia." Journal of Environment and Development 8(2): 152–169.

Sunderlin, William, and J. Rodriguez. 1996. "Cattle, Broadleaf Forests, and the Agricultural Modernization Law of Honduras: The Case of Olancho." Center for International Forestry Research (CIFOR), Occasional Paper #7, Bogor, Indonesia.

Tacconi, L. 2003. "Fires in Indonesia: Causes, Costs, and Policy Implications." Occasional Paper #38, Center for International Forestry Research (CIFOR), Bogor, Indonesia.

Tachibana, Towa, T. Nguyen, and K. Otsuka. 2001. "Agricultural Intensification Versus Extensification: A Case Study of Deforestation in the Northern Hill Region of Vietnam." *Journal of Environmental Economics and Management* 41(1): 44–69.

Tekle, Kebrom, and L. Hedlund. 2000. "Land Cover Changes Between 1958 and 1986 in Kalu District, Southern Wello, Ethiopia." *Mountain Research and Development* 20(1): 42–51.

Terborgh, John. 1999. *Requiem for Nature*. Washington, D.C.: Island Press.

Thiesenhusen, William C. 1991. "Implications of the Rural Land Tenure System for the Environmental Debate: Three Scenarios." *Journal of Developing Areas* 26: 1–24.

Thomsen, Jorgen, C. Mitchell, R. Piland, and J. Donnaway. 2001. "Monitoring Impacts of Hydrocarbon Exploration in Sensitive Terrestrial Ecosystems: Perspectives from Block 78, Peru." Pp. 90–112 in I. Bowles and G. Prickett (eds.). *Footprints in the Jungle: Natural Resource Industries, Infrastructure, and Biodiversity Conservation*. New York: Oxford University Press.

Tiffen, Mary, M. Mortimore, and F. Gichuki. 1994. *More People, Less Erosion: Environmental Recovery in Kenya*. New York: John Wiley and Sons.

Tole, Lise. 2002. "Habitat Loss and Anthropogenic Disturbance in Jamaica's Hellshire Hills Area." *Biodiversity and Conservation* 11: 575–598.

Tucker, Richard P. 2000. *Insatiable Appetite: The United States and the Ecological Degradation of the Third World*. Berkeley, Calif.: University of California Press.

Tugwell, Guy Rexford. 1968 (1946). *This Stricken Land: The Story of Puerto Rico*. New York: Greenwood Press.

Turner, B. L., II, S. Villar, D. Foster, J. Geoghegan, E. Keys, P. Klepeis, D. Lawrence, P. Mendoza, S. Manson, Y. Ogneva-Himmelberger, A. Plotkin, D. Salicrup, R. Choudhury, B. Savitsky, L. Schneider, B. Schmook, and C. Vance. 2001. "Deforestation in the Southern Yucatan Peninsular Region: An Integrative Approach." *Forest Ecology and Management* 154: 353–370.

Turner, M. D. 1999. "Merging Local and Regional Analyses of Land-Use Change: The Case of Livestock in the Sahel." *Annals of the Association of American Geographers* 89(2): 191–219.

Uhlig, Harold. 1988. "Spontaneous and Planned Settlement in Southeast Asia." Pp. 7–43 in Wm. Manshard and W. Morgan (eds.). *Agricultural Expansion and Pioneer Settlement in the Humid Tropics*. Tokyo: United Nations University Press.

Uphoff, Norman. 2001. "Balancing Development and Environment Goals

Through Community Based Natural Resource Management." Pp. 433–450 in D. R. Lee and C. B. Barrett (eds.). *Tradeoffs or Synergies? Agricultural Intensification, Economic Development, and the Environment.* New York: CAB International.

Utting, Peter. 1993. *Trees, People, and Power: Social Dimensions of Deforestation and Forest Protection in Central America.* London: Earthscan.

Vandermeer, John. 1996. "The Human Niche and Rain Forest Preservation in Southern Central America." Pp. 216–229 in Leslie Sponsel, T. Headland, and R. Bailey (eds.). *Tropical Deforestation: The Human Dimension.* New York: Columbia University Press.

Vandermeer, John, and Ivette Perfecto. 1995. *Breakfast of Biodiversity: The Truth about Rain Forest Destruction.* Oakland, Calif.: Institute for Food and Development Policy.

Varughese, George. 2000. "Population and Forest Dynamics in the Hills of Nepal: Institutional Remedies by Rural Communities." Pp. 193–226 in C. Gibson, M. McKean, and E. Ostrom (eds.). *People and Forests: Communities, Institutions, and Governance.* Cambridge, Mass.: MIT Press.

Vayda, A. P. 1983. "Progressive Contextualization: Methods for Research in Human Ecology." *Human Ecology* 11(3): 265–281.

Vayda, A. P., and A. Sahur. 1985. "Forest Clearing and Pepper Farming by Bugis Migrants in East Kalimantan: Antecedents and Impact." *Indonesia* 39: 93–110.

Vayda, A. P., and B. Walters. 1999. "Against Political Ecology." *Human Ecology* 27(1): 167–179.

Vedder, Amy, L. Naughton-Treves, A. Plumptre, L. Mabulana, R. Rutagarama, and W. Weber. 2001. "Epilogue: Conflict and Cooperation in the African Rain Forests." Pp. 557–562 in W. Weber, L. White, A. Vedder, and L. Naughton-Treves (eds.). *African Rain Forest Ecology and Conservation.* New Haven, Conn.: Yale University Press.

Verissimo, A., M. Cochrane, and C. Souza. 2002. "National Forests in the Amazon." *Science* 297: 1478.

von Thunen, Johann H. 1875. *The Isolated State in Relation to Land Use and National Economy.* Berlin: Schmaucher Zarchlin.

Vosti, Stephen A., C. Carpentier, J. Witcover, and J. Valentim. 2001. "Intensified Small-Scale Livestock Systems in the Western Brazilian Amazon." Pp. 113–134 in A. Angelsen and D. Kaimowitz (eds.). *Agricultural Technologies and Tropical Deforestation.* New York: CAB International.

Vosti, Stephen A., J. Witcover, and C. Carpentier. 2002. "Agricultural Intensification by Smallholders in the Western Brazilian Amazon: from Deforestation to Sustainable Land Use." Research Report #130. Washington, D.C.: International Food Policy Research Institute.

Wagner, Michael, and J. Cobbinah. 1993. "Deforestation and Sustainability in Ghana: The Role of Tropical Forests." *Journal of Forestry* 91(6): 35–39.

Walker, Robert T. 1987. "Land Use Transition and Deforestation in Developing Countries." *Geographical Analysis* 19: 8–30.

————. 2003. "Mapping Process to Pattern in the Landscape Change of the Amazonian Frontier." *Annals of the Association of American Geographers* 93(2): 376–398.

Wallace, Alfred Russel. 1962 (1869). *The Malay Archipelago.* New York: Dover.

Walsh, P. D., K. Abernethy, M. Bermejo, R. Beyersk, P. De Wachter, M. Akou, B. Huljbregis, D. Mambounga, A. Toham, M. Kilbourn, S. Lahm, S. Latour, F. Maisels, C. Mbina, Y. Mihindou, S. Obiang, E. Effa, M. Starkey, P. Telfer, M. Thibault, C. Tutin, L. White, and D. Wilkie. 2003. "Catastrophic Ape Decline in Western Equatorial Africa." *Nature* 422(6932): 611–614.

Walters, B. B. 2003. "People and Mangroves in the Philippines: Fifty Years of Coastal Environmental Change." *Environmental Conservation* 30(2): 293–303.

Watson, Mary K. 1978. "The Scale Problem in Human Geography." *Geografiska Annaler* 60B: 36–47.

Weil, Connie, and J. Weil. 1983. "Government, Campesinos, and Business in the Chapare: A Case Study of Amazonian Occupation." *Inter-American Economic Affairs* 36(4): 29–62.

Weis, Tony. 2000. "Beyond Peasant Deforestation: Environment and Development in Rural Jamaica." *Global Environmental Change* 10: 299–305.

Whitmore, T. C. 1995. "Comparing Southeast Asian and Other Tropical Rainforests." Pp. 5–15 in R. Primack and T. Lovejoy (eds.). *Ecology, Conservation, and Management of Southeast Asian Rain Forests.* New Haven, Conn.: Yale University Press.

Wickramagamage, P. 1998. "Large-Scale Deforestation for Plantation Agriculture in the Hill Country of Sri Lanka and Its Impacts." *Hydrological Processes* 12: 2015–2028.

Wilkie, David. 1996. "Logging in the Congo: Implications for Indigenous Foragers and Farmers." Pp. 230–247 in Leslie Sponsel, Thomas Headland, and Robert C. Bailey (eds.). *Tropical Deforestation: The Human Dimension.* New York: Columbia University Press.

Wilkie, David, J. Carpenter, and Q. Zhang. 2001. "The Under Financing of Protected Areas in the Congo Basin: So Many Parks and So Little Willingness to Pay." *Biodiversity and Conservation* 10(5): 691–709.

Wilkie, David, and J. Finn. 1988. "A Spatial Model of Land Use and Forest Regeneration in the Ituri Forest of Northeastern Zaire." *Ecological Modelling* 41: 307–323.

Wilkie, David S., J. Sidle, and G. Boundzanga. 1992. "Mechanized Logging, Market Hunting, and a Bank Loan in Congo." *Conservation Biology* 6(4): 1–11.

Wilks, Ivor. 1978. "Land, Labour, Capital, and the Forest Kingdom of Asante: A Model of Early Change." Pp. 487–534 in J. Friedman and M. J. Rowland (eds.). *The Evolution of Social Systems.* Pittsburgh, Penn.: University of Pittsburgh Press.

Williams, Robert G. 1986. *Export Agriculture and the Crisis in Central America*. Chapel Hill, N.C.: University of North Carolina Press.

Wilshusen, Peter, S. Brechin, C. Fortwangler, and P. West. 2002. "Reinventing a Square Wheel: Critique of a Resurgent Protection Paradigm in International Biodiversity Conservation." *Society and Natural Resources* 15(1): 17–40.

Wily, Liz A. 2002. "The Political Economy of Community Forestry in Africa: Getting the Power Relations Right." *Forests, Trees, and People* #46. Umeå, Sweden: Swedish University of Agricultural Sciences.

Witte, John. 1992. "Deforestation in Zaire: Logging and Landlessness." *The Ecologist* 22(2): 58–64.

World Bank. 2001. "Data and Statistics." Accessed at www/worldbank.org/data.

World Wide Fund for Nature, 1999, "Slip Sliding Away …," *Arborvitae: The IUCN/WWF Forest Conservation Newsletter* #12. Gland, Switzerland.

Wrong, Michela. 2001. *In the Footsteps of Mr. Kurtz: Living on the Brink of Disaster in Mobutu's Congo*. New York: Harper Collins.

Wunder, Sven. 2000. *The Economics of Deforestation: The Case of Ecuador*. London: MacMillan.

———. 2001. "Ecuador Goes Bananas: Incremental Technological Change and Forest Loss." Pp. 167–194 in A. Angelsen and D. Kaimowitz (eds.). *Agricultural Technologies and Tropical Deforestation*. New York: CAB International.

———. 2003. *Oil Wealth and the Fate of the Forests*. London: Routledge.

Zweifler, Michael, M. Gold, and R. Thomas. 1994. "Land Use Evolution in Hill Regions of the Dominican Republic." *Professional Geographer* 46(1): 39–53.

Index

Abidjan (Ivory Coast), 76, 85
Acacia species, 106, 111
Accra (Ghana), 85
afforestation, 161; in Central America and Caribbean, 39, 45, 159, 189n6b; in China, 133, 137, 193n3a, 194n1; in East Africa, 111–15, 118, 120–21, 171; in South Asia, 16, 125, 128–38, 159, 170, 171, 189n6b, 193n3a; in Southeast Asia, 149–54, 162, 171, 189n6b, 193nn1,2b; in West Africa, 83, 88, 171, 190n1b, 191n4. *See also* reforestation; secondary growth
Africa: forest cover change in, 6, 25, 125, 187n3; and peri-urban deforestation, 92, 96–98, 101–2, 107, 112–13, 119–20, 159, 160; woodlands of, 5, 91, 105–7, 109, 192n3a. *See also* Central Africa; East Africa; West Africa
agriculture: and certified organic foods, 89; and citemene system, 107, 111, 118–19; and *colono* system, 60–67; and contract farming, 41–44; and

fertilizer use, 35, 41–42, 84, 118; and follow-on farmers, 80, 99, 153; and food crops, 45, 77, 85, 87, 97; and fruit trees, 80, 83, 88, 115, 151; markets for, 14–15, 18–19, 35–36, 39–40; and sedentary cultivators, 145, 149, 152, 191n6; slash-and-burn, 106; swidden, 145. *See also* shifting cultivators; smallholders
agro-forestry: in Africa, 111, 120–21, 191n7; in Brazil, 61, 64, 65–68, 189n2; and community-based management, 16–17, 88, 119–22, 126–27, 130–38, 163–65, 194n3; in Costa Rica, 42–44, 46; in Ecuador, 27, 28, 42, 57, 164, 191n6; markets for, 3, 15, 18–19, 25, 65–68, 142, 153; in Southeast Asia, 146–48, 150, 154; and wood value, 13–15, 17, 128, 130, 150
Akwapim people (West Africa), 78
Albert National Park (Rwanda), 116
Albertine rift, 91, 106
Altamira (Para), 66, 68
Amanor, Kojo, 84